Strongly Stabilizable Distributed Parameter Systems

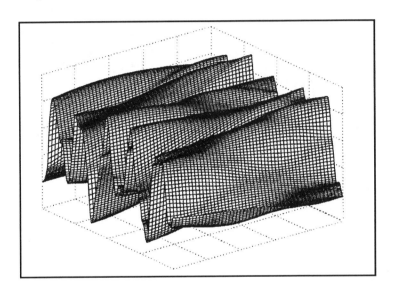

FRONTIERS IN APPLIED MATHEMATICS

The SIAM series on Frontiers in Applied Mathematics publishes monographs dealing with creative work in a substantive field involving applied mathematics or scientific computation. All works focus on emerging or rapidly developing research areas that report on new techniques to solve mainstream problems in science or engineering.

The goal of the series is to promote, through short, inexpensive, expertly written monographs, cutting edge research poised to have a substantial impact on the solutions of problems that advance science and technology. The volumes encompass a broad spectrum of topics important to the applied mathematical areas of education, government, and industry.

EDITORIAL BOARD

BOOKS PUBLISHED IN FRONTIERS
IN APPLIED MATHEMATICS

Oostveen, Job, *Strongly Stabilizable Distributed Parameter Systems*

Griewank, Andreas, *Evaluating Derivatives: Principles and Techniques of Algorithmic Differentiation*

Kelley, C. T., *Iterative Methods for Optimization*

Greenbaum, Anne, *Iterative Methods for Solving Linear Systems*

Kelley, C. T., *Iterative Methods for Linear and Nonlinear Equations*

Bank, Randolph E., *PLTMG: A Software Package for Solving Elliptic Partial Differential Equations. Users' Guide 7.0*

Moré, Jorge J. and Wright, Stephen J., *Optimization Software Guide*

Rüde, Ulrich, *Mathematical and Computational Techniques for Multilevel Adaptive Methods*

Cook, L. Pamela, *Transonic Aerodynamics: Problems in Asymptotic Theory*

Banks, H. T., *Control and Estimation in Distributed Parameter Systems*

Van Loan, Charles, *Computational Frameworks for the Fast Fourier Transform*

Van Huffel, Sabine and Vandewalle, Joos, *The Total Least Squares Problem: Computational Aspects and Analysis*

Castillo, José E., *Mathematical Aspects of Numerical Grid Generation*

Bank, R. E., *PLTMG: A Software Package for Solving Elliptic Partial Differential Equations. Users' Guide 6.0*

McCormick, Stephen F., *Multilevel Adaptive Methods for Partial Differential Equations*

Grossman, Robert, *Symbolic Computation: Applications to Scientific Computing*

Coleman, Thomas F. and Van Loan, Charles, *Handbook for Matrix Computations*

McCormick, Stephen F., *Multigrid Methods*

Buckmaster, John D., *The Mathematics of Combustion*

Ewing, Richard E., *The Mathematics of Reservoir Simulation*

Strongly Stabilizable Distributed Parameter Systems

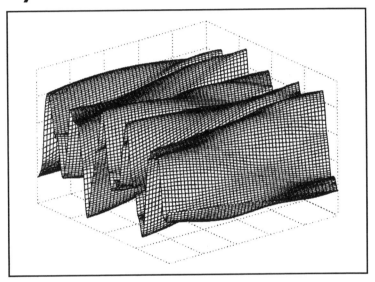

Job Oostveen

Philips Research Laboratories
Eindhoven, The Netherlands

Society for Industrial and Applied Mathematics
Philadelphia

Library of Congress Cataloging-in-Publication Data

Oostveen, Job.
    Strongly stabilizable distributed parameter systems / Job Oostveen.
        p. cm. — (Frontiers in applied mathematics ; 20)
    Includes bibliographical references and index.
    ISBN 0-89871-455-9
        1. Distributed parameter systems. 2. Stability. I. Title. II. Series.

QA402 .O58 2000
003'.78—dc21

00-026286

 is a registered trademark.

Contents

Preface

The theory of distributed parameter systems is becoming more and more a complete theory, containing all of the topics that are well known from finite-dimensional systems theory. However, almost without exception, the theory assumes that the systems are either exponentially stable or exponentially stabilizable. This is rather unfortunate, because there are many models of physically relevant systems that are not exponentially stabilizable. Still, some of these models can be made asymptotically stable in some sense. Most importantly, this is the case for models of weakly damped flexible structures and of acoustical problems. In the literature, there exist many articles containing stability analysis and stabilization methods for such engineering systems. However, unlike the case of exponentially stabilizable systems, until recently there was not a well-structured theory of how to deal with such systems. To be able to deal with such models, ad-hoc techniques had to be applied, very often based on knowledge of a particular Lyapunov function. Moreover, the convergence properties of the approximation methods used were unknown.

In this book, we try to fill this gap and to provide a framework for a structured approach to strongly stabilizable systems. Most importantly, we present extensions of system theoretic notions and results to the case of strong stability. We believe that this framework will be of use to people working with strongly stabilizable systems, as they will become less dependent on ad-hoc methods.

The choice of topics in this book is strongly influenced by the book *An Introduction to Infinite-Dimensional Linear Systems* by Curtain and Zwart [36]. Moreover, in the same way as [36], our book uses the state-space and frequency domain approach in an integrated fashion. It represents the current status of research for strongly stabilizable systems. We discuss possible definitions of strong stability, stabilizability, and detectability. Moreover, we develop a Riccati theory including numerical approximation of strongly stabilizing solutions and we discuss state-space formulas for (normalized) coprime factorizations. We discuss possibilities of input-output stabilization by dynamic compensators. The robustness of stability is studied, with respect both to nonlinear perturbations and to coprime factor perturbations. We have tried to motivate the results, as much as possible, by supplying a number of models of physical/engineering systems that satisfy the assumptions of our theory. All the theory in this book is restricted to bounded input and output operators. Although this excludes many important examples, there are still many motivating physical systems that do satisfy the assumptions of our theory. These include certain examples of boundary control systems.

This book is intended for people working on the analysis and control of systems described by weakly damped partial differential equations, for instance, related to the stabilization of weakly damped flexible structures or to active noise control. Also, it will be of interest to system theorists with an interest in infinite-dimensional systems.

This book is a slightly adapted and extended version of my Ph.D. thesis. As such it represents the outcome of a period of four years at the Department of Mathematics of the University of Groningen, the Netherlands. My supervisor during this period was Ruth Curtain. All results in this thesis are the product of cooperation with her and guidance by her. I would like to express my deepest gratitude for this. I thank Tom Banks for bringing the manuscript to the attention of SIAM and for his invitation for a very inspiring three month stay at his Center for Research in Scientific Computing at North Carolina State University. I have benefitted much from the questions, comments, and suggestions of a number of colleagues, for which I am very grateful. Frank Callier read the entire manuscript of my thesis and came up with an enormous number of corrections and suggestions for improvements. We worked on the results in Chapter 5 with Kazi Ito, whose persistent questions also had a large impact on Chapter 3. Finally, I am grateful for the pleasant cooperation of Marianne Will and her colleagues at SIAM.

Job Oostveen
Eindhoven, 1999

Notation

Symbol	Meaning	Page
$*$	A^*, adjoint operator of A	
$\bar{s}$	complex conjugate of s	
$\sim$	$G^\sim(s) = G(-\bar{s})^*$	33
$\underset{\tau}{\diamond}$	$f_1 \underset{\tau}{\diamond} f_2$, concatenation of f_1 and f_2 at τ	35
$\lVert \cdot \rVert$	norm	
$\lVert \cdot \rVert_Z$	norm in space Z	
$\lVert \cdot \rVert_2$	$\mathbf{H}_2$- or $\mathbf{L}_2$-norm	7
$\lVert \cdot \rVert_\infty$	$\mathbf{H}_\infty$- or $\mathbf{L}_\infty$-norm	7
$\langle \cdot, \cdot \rangle$	inner product	
$\langle \cdot, \cdot \rangle_Z$	inner product in space Z	

Roman

Symbol	Meaning	Page
$\mathbb{C}_0^+$	$\{s \in \mathbb{C} \mid \mathrm{Re}(s) > 0\}$	
$\bar{\mathbb{C}}_0^+$	$\{s \in \mathbb{C} \mid \mathrm{Re}(s) \geq 0\}$	
$\mathrm{col}(a,b)$	$\begin{pmatrix} a \\ b \end{pmatrix}$	
$D(A)$	domain of operator A	
$\mathbb{F}_t$	input-output map	10, 35
$\mathbb{F}$	extended input-output map	11
$\mathbf{F}$	family of input-output maps	35
H_G	Hankel operator with symbol G	87
$\mathbf{H}_\infty(B)$	Hardy space of bounded, holomorphic functions on $\mathbb{C}_0^+$ with values in B	8
$\mathbf{H}_\infty$	$\mathbf{H}_\infty(B)$, where B is not specified	
$\mathbf{H}_\infty[\mathbf{H}_\infty]^{-1}$	quotient field of $\mathbf{H}_\infty$	96
$\mathbf{H}_2(B)$	Hardy space of square integrable functions on $\mathbb{C}_0^+$ with values in B	7
$H^n(a,b)$	space of real-valued functions on the interval (a,b), of which the derivatives of order $0,\dots,n$ are in $\mathbf{L}_2(a,b)$	116
j	imaginary unit	

$\mathcal{L}(Z_1, Z_2)$	Banach space of bounded operators from Z_1 to Z_2	
$\mathcal{L}(Z)$	$\mathcal{L}(Z, Z)$	
$\mathbf{L}_2(a, b; Z)$	class of Lebesgue measurable Z-valued functions, with $\int_a^b \|f(t)\|^2 dt < \infty$	7
$\mathbf{L}_2^{loc}(0, \infty; Z)$	class of functions which are in $\mathbf{L}_2(0, T; Z)$ for all $T \geq 0$	7
$\mathbf{L}_\infty(a, b; Z)$	class of bounded Lebesgue measurable Z-valued functions on (a, b)	7
$\mathbf{L}_\infty(Z)$	$\mathbf{L}_\infty(-j\infty, j\infty; Z)$	7
$\mathcal{M}A$	class of matrices with elements in A	96
$\mathrm{Re}(s)$	real part of complex number s	
$r(A)$	spectral radius of operator A	
$\mathbf{T}$	C_0-semigroup	35
$\mathcal{T}_\Pi$	Toeplitz operator with symbol Π	33
$\mathcal{T}_\Pi^t$	time-domain Toeplitz operator with symbol Π	33

Greek

Γ	time-domain Hankel operator	88
$\rho(A)$	resolvent set of operator A	
$\Sigma(A, B, C, D)$	state-space system	10
$\Sigma(A, B, -)$	state-space system, in the case C is irrelevant	
Φ_t	input map	10
$\tilde{\Phi}$	extended input map	10
$\mathbf{\Phi}$	family of input maps	35
Ψ_t	output map	10
$\tilde{\Psi}$	extended output map	10
$\mathbf{\Psi}$	family of output maps	35

Chapter 1

Introduction

1.1 Motivation

Questions about stability arise in almost every control problem. As a consequence, stability is one of the most extensively studied subjects in systems theory, and a great diversity of stability notions has been introduced. Many of the prevailing stability definitions are concerned with autonomous differential equations, which in the linear time-invariant case have the form

$$\dot{z}(t) = Az(t), \quad z(0) = z_0 \in Z.$$

When $Z = \mathbb{R}^n$, such systems correspond to ordinary differential equations and all solutions are given by $z(t) = e^{At}z_0$. In the case that Z is an infinite-dimensional space (we will consider only the case that Z is a Hilbert space), A is a possibly unbounded operator on Z and solutions are given by $z(t) = T(t)z_0$, where $T(t)$ is the C_0-semigroup generated by A (a C_0-semigroup is the generalization of e^{At} to the case of unbounded operators A). Many linear partial differential equations and delay-differential equations can be formulated in this way as first order equations on appropriately chosen Hilbert spaces. For such autonomous systems, by *stability* we mean global asymptotic stability of the origin; i.e., $z(t)$ tends to zero as t tends to infinity for all initial conditions $z_0 \in Z$. If a finite-dimensional system is globally asymptotically stable, you can even say more: It is *exponentially stable*; i.e., there exist constants $\alpha > 0$ and $M \geq 1$ such that for all z_0,

$$\|z(t)\| = \|e^{At}z_0\| \leq Me^{-\alpha t}\|z_0\|.$$

This formula expresses the fact that for exponentially stable systems, there is a uniform bound α on the rate of decay of all trajectories.

There are many possible stability definitions for the infinite-dimensional case. In systems theory, *exponential stability* is the most desirable type of stability. This means that, as for finite-dimensional systems, there is a uniform rate of decay of solutions: there exist constants $\alpha > 0$ and $M \geq 1$ such that for all z_0,

$$\|z(t)\| = \|T(t)z_0\| \leq Me^{-\alpha t}\|z_0\|.$$

Virtually all results in infinite-dimensional systems theory have used exponential stability. There are, however, partial differential equations that are globally asymptotically stable, but whose solutions do not have a uniform rate of decay. Even if a uniform bound on the rate of decay exists, it may be very hard to find. Thus, there is a need to have stability definitions that do not assume a uniform rate of decay on the solutions. The following example from Russell [91] is a system whose trajectories all converge to zero, but for which there is no uniform rate of decay.

Example 1.1.1 The following model describes the displacement in a weakly damped vibrating string of length 1, with clamped ends. The displacement $y(x,t)$ of the string at position x and time t satisfies

$$\frac{\partial^2 y}{\partial t^2}(x,t) + D\frac{\partial y}{\partial t}(x,t) = \frac{\partial^2 y}{\partial x^2}(x,t) \quad \text{for } t \geq 0 \text{ and } 0 \leq x \leq 1, \tag{1.1}$$

with boundary conditions $y(0,t) = y(1,t) = 0$. Here, D is a damping operator, which is defined below. For easy reference, we define the operator A by $Ay = -\partial^2 y/\partial x^2$, with domain of definition $\mathcal{D}(A) = \{y \in H^2(0,1) | \ y(0) = y(1) = 1\}$ ($H^2(0,1)$ is the space of functions $y \in \mathbf{L}_2(0,1)$ that are twice differentiable almost everywhere on $(0,1)$ and whose first- and second-order derivatives are in $\mathbf{L}_2(0,1)$). A has eigenvalues $\lambda_k = k^2\pi^2$, $k = 1,2,\ldots$, with eigenvectors $\phi_k(x) = \sqrt{2}\sin(k\pi x)$ forming a complete orthonormal set in $\mathbf{L}_2(0,1)$.

D is defined by $Dy = \varepsilon\langle g,y\rangle g$, where $\varepsilon > 0$ and

$$g = \sum_{k=1}^{\infty} \gamma_k\phi_k,$$

with γ_k satisfying $0 < |\gamma_k| \leq M/\sqrt{\lambda_k}$. For example, $\gamma_k = 1/k$ would satisfy this constraint. Next, we define the Hilbert space $Z = \mathbf{L}_2(0,1) \times \mathbf{L}_2(0,1)$. Introducing

$$z(t) = \begin{pmatrix} A^{\frac{1}{2}}y(x,t) \\ \frac{\partial y}{\partial t}(x,t) \end{pmatrix} \quad \text{and} \quad \mathcal{A} = \begin{bmatrix} 0 & A^{\frac{1}{2}} \\ -A^{\frac{1}{2}} & -D \end{bmatrix},$$

(1.1) can be rewritten as

$$\dot{z}(t) = \mathcal{A}z(t), \ z(0) = z_0 = \begin{pmatrix} A^{\frac{1}{2}}y(x,0) \\ \frac{\partial y}{\partial t}(x,0) \end{pmatrix}.$$

All solutions are given by $z(t) = T(t)z_0$, where $T(t)$ is the semigroup generated by $\mathcal{A}$ on Z. Russell has shown that all solutions converge to zero. Furthermore, he showed that all eigenvalues of $\mathcal{A}$ have negative real part and are given by

$$\nu_k = ik\pi - \frac{\varepsilon}{2}\gamma_{|k|}^2 + \mathcal{O}\left(\frac{\varepsilon^2}{k^2\pi^2}\right).$$

The corresponding eigenvectors ψ_k form a Riesz basis of Z. If we take the initial condition $z_0 = \psi_k$, the solution is given by $z(t) = e^{\nu_k t}\psi_k$ and this converges to zero as t tends to infinity. Because the real part of ν_k converges to zero as k tends to infinity, there is no uniform bound on the rate of decay. The system is therefore globally asymptotically stable, but *not* exponentially stable.   ◇

This example shows that for partial differential equations the concept of exponential stability is not always applicable. Systems theory for such models should therefore incorporate forms of stability weaker than exponential stability. One such form of stability is *strong stability*: a semigroup $T(t)$ is strongly stable if all trajectories converge to zero, i.e.,

$$\lim_{t\to\infty} \|T(t)z\| = 0 \quad \text{for all } z \in Z.$$

The semigroup occurring in Example 1.1.1 is an instance of a strongly stable semigroup. If the trajectories only converge weakly to zero, the semigroup is called *weakly stable*:

$$\lim_{t\to\infty} \langle T(t)z_1, z_2 \rangle = 0 \text{ for all } z_1, z_2 \in Z.$$

In this book we concentrate on developing system theoretic results and control theoretic methods for infinite-dimensional systems in the context of *strong stability*.

In systems theory, one does not study autonomous systems, but systems with inputs and outputs. In the linear time-invariant case, these have the form

$$\dot{z}(t) = Az(t) + Bu(t), \quad z(0) = z_0,$$
$$y(t) = Cz(t) + Du(t).$$

Again, A, B, C, and D are matrices in the case of a finite-dimensional system and operators in the case of infinite-dimensional systems. In what follows, we will call such systems "control systems" to distinguish them from the "autonomous systems" that we discussed first. Clearly, the different types of asymptotic stability that were defined above for autonomous systems are of importance for control systems, as well, since they concern the solutions with input $u(t) = 0$. Stability notions for control systems that deal with solutions for $u = 0$ are called *internal stability*. However, there is a different class of important notions of stability for control systems, called *external stability*. These are concerned with the effects of inputs and the effects on outputs. The most important notion is that of input-output stability: We call a system *input-output stable* if there exists a finite number $\gamma > 0$ such that all solutions $(u(t), z(t), y(t))$ with initial condition $z_0 = 0$ satisfy

$$\|y(t)\|_{L_2(0,\infty;Y)} \leq \gamma \|u(t)\|_{L_2(0,\infty;U)}.$$

If the semigroup of an infinite-dimensional system is exponentially stable, then the system is automatically also input-output stable. Many results in systems theory make use of this implication. For a strongly stable semigroup it does not hold, unfortunately, and this is the cause of a number of complications in the construction of a systems theory for strongly stabilizable systems. It has become clear that calling a control system *strongly stable* if its semigroup is strongly stable is not an appropriate definition for arriving at useful system theoretic results. This leads us to incorporate certain external stability properties into the definition of a strongly stable system.

It is well known that a finite-dimensional linear system is stable if all the eigenvalues of the system matrix A (or all the roots of the characteristic polynomial) are in the left half-plane. Thus, stability can be tested by algebraic methods. For the strong stability of a semigroup, such a simple characterization in terms of spectral properties of its generator A does not exist. Therefore, different, analytical tools must be used to test whether a system is strongly stable. As for nonlinear systems, a Lyapunov analysis can be effective. In the case of physical systems, there is often an obvious choice for a candidate Lyapunov function: the energy of the system. This energy-based approach leads us into the theory of passive systems. Passivity is a notion originating from network theory. Let us first give an example.

Example 1.1.2 Consider the model for the motion of a mass-spring-damper system. The displacement of the mass from the equilibrium position is denoted $x(t)$ and an external force $u(t)$ acts on the mass:

$$m\ddot{x}(t) + d\dot{x}(t) + kx(t) = u(t),$$
$$y(t) = \dot{x}(t).$$

We take $z(t) = (x(t), \dot{x}(t))$ as the state of this system. The total energy in the system is $E(z(t)) = \frac{1}{2}kx(t)^2 + \frac{1}{2}m\dot{x}(t)^2$, and an easy calculation shows that the change of energy satisfies

$$E(z(t)) - E(z(0)) = \int_0^t \left[-d\dot{x}(s)^2 + \dot{x}(s)u(s) \right] ds \leq \int_0^t \dot{x}(s)u(s)ds$$

$$= \int_0^t y(s)u(s)ds.$$

This inequality expresses the fact that the increase of energy stored in the system does not exceed the work applied by the external force (i.e., it does not exceed the energy supplied by the external source). $\diamond$

Passivity is an abstract version of this physical principle. A control system is called passive if there exists a function $S : Z \to \mathbb{R}^+$ (the storage function) such that the *dissipation inequality*

$$S(z(t)) - S(z(0)) \leq \int_0^t \langle u(s), y(s) \rangle ds$$

is satisfied. As is well known, passivity of linear systems is closely related to the system theoretic property of positive real transfer functions. In this book we will pay special attention to passive/positive real systems, and in many instances we can use the special structure of these systems to obtain sharper results.

The initial motivation for the research described in this book was to develop a systems theory for a class of systems occurring as models of flexible structures. These are often modelled as linear systems on a Hilbert space. When the norm is chosen such that it corresponds to the energy of the system, the passive structure is especially highlighted. In this case, they have a state-space representation $\Sigma(A, B, B^*)$, where A generates a contraction semigroup. The systems we are interested in here are dissipative systems with collocated sensors and actuators. The term "dissipative" occurs because the generator A of a contraction semigroup is a dissipative operator. For a long time, it has been known by engineers that a partial differential equation describing a mechanical system, such as a flexible structure, leads to a positive real system if actuators and sensors are designed in a "collocated" fashion. This means that measurements and control action are made dual in some sense; a necessary condition for this is that actuators and sensors are implemented at the same location. If an actuator exerts a force at a particular location, a sensor is collocated with this actuator if it measures the linear velocity at the same point. Example 1.1.2 is precisely such an example with collocated actuator and sensor. This is made more precise in Example 2.2.5. Similarly, if the actuator exerts a moment, the sensor must measure the angular velocity at the same location. Having designed collocated actuators and sensors for a dissipative mechanical structure, and thus having obtained a positive real system, negative feedback (i.e., $u = -y$) stabilizes the system in an input-output sense. This knowledge was used to control flexible structures and the method works even if no detailed model of the system is available, as long as the passive structure is present.

In this book, we use the term "dissipative" for operators A that satisfy the relation $\langle Az, z \rangle + \langle z, Az \rangle \leq 0$ for all $z \in \mathcal{D}(A)$ (see Definition 2.2.1). To avoid confusion, we would like to stress that we do not use "dissipativity" in its system theoretic or physical sense. In systems theory, dissipativity is a generalization of passivity to allow for different supply rates (the supply rate is the quadratic form in u and y occurring under the integral sign in the dissipation inequality). In this book, we are concerned only with the standard supply rate $\langle u(t), y(t) \rangle$ and thus we use passivity instead of dissipativity. In physics, the term dissipative is used for systems in which energy is dissipated. For linear systems on an appropriately chosen Hilbert space, this implies the dissipativity of the operator A. However, a dissipative operator A can also represent a system in which all energy is conserved. If in what follows we use the terminology "dissipative system," then we refer to systems in our special class $\Sigma(A, B, B^*)$: dissipative systems with collocated actuators and sensors.

During the development of this theory, it appeared that many of the results could be formulated for general strongly stabilizable systems, without the assumption of the special structure of dissipative systems with collocated actuators and sensors. Even though we have chosen to give these more general formulations, the models with the special dissipative structure are still the main motivation for this research. Most of our physical examples in Chapter 9 exhibit this form and for many of our results we have given the (often easier) formulation for dissipative systems with collocated actuators and sensors as a corollary.

1.2 Overview of the Results

The results in this book can be divided into two parts. First, a framework that is appropriate for studying control systems in the context of strong stability is developed. Second, a diversity of results is obtained within this framework. In this section, we give an overview of the results so that the interested reader can choose his or her favorite route through them.

Chapter 2: Two Main Themes

This chapter describes the two central objects of our study. First of all, strong stability, stabilizability and detectability are defined and we derive some important related properties of strongly stable systems. Second, we introduce the class of dissipative systems with collocated actuators and sensors in all its mathematical detail and show the useful properties that it exhibits.

Chapter 3: Stabilizability and Detectability

The definitions of strong stabilizability and strong detectability were introduced in Chapter 2 without any motivation. In Chapter 3, we show that these definitions are indeed appropriate by showing that they can be used to generalize certain well-known system theoretic results to the case of strong stability. These results are

- the equivalence of strong stability and input-output stability for systems that are strongly stabilizable and strongly detectable;

- the existence of a unique stabilizing solution to the LQ Riccati equation for a strongly stabilizable and strongly detectable system.

These results have also been published in [28].

Chapter 4: Riccati Theory

Probably no control theoretic tool has been studied more thoroughly than the Riccati equation. Therefore, a logical topic with which to start the development of systems theory for strongly stabilizable systems is the theory of existence and uniqueness of strongly stabilizing solutions to algebraic Riccati equations. A start at this subject was made in Chapter 3 with results for the standard LQ Riccati equation. In Chapter 4, we study very general Riccati equations for strongly stabilizable systems. The results were published in [80]. It is shown here that a strongly stabilizing solution exists for a Riccati equation with a general (i.e., not necessarily positive) cost operator and associated with a statically stabilizable system if and only if a certain factorization of the Popov function exists. This basic result is used to derive more specific results for both general statically stabilizable systems and dissipative infinite-dimensional systems with collocated measurement and control. This includes a result for the existence of J-spectral factorizations and versions of the positive real lemma and the bounded real lemma.

Chapter 5: Numerical Approximation

If the theory of Chapter 4 is to be used in a practical control design, then one needs to be able to compute numerical approximations to the operator solution of the Riccati equation. It is important to use approximation algorithms with proven convergence properties. In Chapter 5, the convergence of sequences of stabilizing solutions of finite-dimensional algebraic Riccati equations to the strongly stabilizing solution of the standard LQ Riccati equation is studied and proved. This result was obtained in collaboration with Curtain and Ito [83]. In Chapter 9, we illustrate the effectiveness of the algorithm with a numerical example of the LQ Riccati equation for a problem of sound propagation.

Chapter 6: Coprime Factorizations and Compensators

In the foregoing chapters, stabilization by state feedback or by static output feedback has been considered. Of course, it is also important to have available a theory about dynamic compensation. The relation between dynamic stabilization and coprime factorizations is well known, so in this chapter the two are treated jointly. We give state-space formulas for normalized doubly coprime factorizations for a class of strongly stabilizable and detectable systems. Furthermore, we develop results about dynamic (input-output) stabilization of strongly stabilizable systems. We also study whether the resulting closed-loop system is strongly stable. Unlike the results in the rest of this book, we do not obtain a perfect generalization of the finite-dimensional results for the subjects in this chapter. The chapter ends with a positive result about strong stabilization of a system in the $\Sigma(A, B, B^*)$ class by finite-dimensional strictly positive real controllers.

Results about normalized coprime factorizations for systems in the $\Sigma(A, B, B^*)$ class were published in Oostveen and Curtain [81]. The generalization to more general strongly stabilizable and detectable systems appeared in Curtain and Oostveen [29].

Chapter 7: Robust Stabilization

The question of robustness of control systems has been very prominent in control theory in the 1980s and 1990s. For infinite-dimensional systems it is of even greater importance, because LQG and H_∞-controllers are usually also infinite-dimensional. So, for implementation purposes, one will have to approximate these controllers. In this chapter, we give a parameterization of controllers that robustly stabilize a strongly stabilizable plant with respect to coprime factor perturbations. An important step in the solution of this problem is the explicit parameterization of all solutions to a certain Nehari extension problem.

The results on the Nehari problem were published in [27]. The papers [30, 81] concern the parameterization of robustly stabilizing controllers for the $\Sigma(A, B, B^*)$ class and more general strongly stabilizable and detectable systems, respectively.

Chapter 8: Nonlinear Perturbations

In Chapter 8, we deal with a classical robustness problem: the absolute stability problem. A version of the Popov criterion is presented as a solution. We show that a strongly stable system with strictly positive real transfer function remains asymptotically stable under a class of nonlinear perturbations in the feedback loop. This result is published in [82] and a preliminary version appeared in [26].

Chapter 9: Examples of Strongly Stabilizable Systems

Four examples of partial differential equation models of engineering systems are described in Chapter 9. We show how they are cast into state-space form, in which three of them exhibit the special dissipative structure; so, in particular, they are examples of strongly stabilizable systems. For one of the examples we work out the numerical approximation procedure for the strongly stabilizing solution of the standard LQ Riccati equation, which was described in Chapter 5.

Chapter 10: Conclusions

In this last chapter, we summarize the research described in this book and we discuss some open questions and possibilities for future research in the area of strongly stabilizable infinite-dimensional systems.

Notation

The relevant notation is summarized in a list on pages xi and xii. All notation is defined at the location where it is first used. As an exception to this rule, we give here definitions of the function spaces used for systems and signals.

Let Z be a Hilbert space and B a Banach space. Then

$\mathbf{L}_2(a, b; Z)$ is the space of weakly measurable, Z-valued functions on (a, b), such that $\|f\|_2 = (\int_a^b \|f(t)\|_Z^2 dt)^{\frac{1}{2}} < \infty$;

$\mathbf{L}_2(Z)$ is the space of weakly measurable, Z-valued functions defined on the imaginary axis, such that $\|f\|_2 = (\int_{-\infty}^{\infty} \|f(j\omega)\|_Z^2 d\omega)^{\frac{1}{2}} < \infty$;

$\mathbf{L}_2^{loc}(0, \infty; Z)$ is the Fréchet space of weakly measurable, Z-valued functions that are in $\mathbf{L}_2(0, T; Z)$ for all $T \geq 0$;

$\mathbf{L}_\infty(a, b; B)$ is the space of weakly measurable, B-valued, essentially bounded functions, defined on (a, b);

$\mathbf{L}_\infty(B)$ is the space of weakly measurable, B-valued, essentially bounded functions, defined on the imaginary axis, with norm $\|f\|_\infty = \text{ess sup}_{\omega \in \mathbb{R}} \|f(j\omega)\|_\infty$;

$\mathbf{H}_2(Z)$ is the space of holomorphic, Z-valued functions defined on the open right half-plane, such that $\|f\|_2 = \sup_{x > 0} (\int_{-\infty}^{\infty} \|f(x + jy)\|_Z^2 dy)^{\frac{1}{2}} < \infty$;

$\mathbf{H}_2^\perp(Z)$ is the left half-plane counterpart of $\mathbf{H}_2(Z)$.

$\mathbf{H}_\infty(B)$ is the space of holomorphic, B-valued functions defined on the open right-half plane, such that $\|f\|_\infty = \sup_{x>0} \|f(x+jy)\|_Z < \infty$.

Note that $\|\cdot\|_\infty$ denotes the norm both in $\mathbf{L}_\infty$ and $\mathbf{H}_\infty$. This does not cause any problems, because if a function f is in $\mathbf{H}_\infty$, then it has an extension to the imaginary axis and its $\mathbf{H}_\infty$-norm equals the $\mathbf{L}_\infty$-norm of its boundary function $f(j\omega)$. A similar remark applies to the norms of $\mathbf{L}_2$ and $\mathbf{H}_2$.

In the above definitions, all types of $\mathbf{L}_2$- and $\mathbf{H}_2$-spaces are Hilbert spaces. In Chapter 7, we also use $\mathbf{H}_2(B)$, with B a finite-dimensional Banach space. It is defined in the same way as $\mathbf{H}_2(Z)$, above. However, because B has no inner product, $\mathbf{H}_2(B)$ has no inner product either, and it is not a Hilbert space.

Chapter 2

Two Main Themes

This book is centered around two main themes. The first one is strong stability. As we have explained in the introduction, it is not always possible to work with the notion of exponential stability. Therefore, we define the weaker notion of strong stability for control systems in Section 2.1. In Section 2.2, we consider the second main theme: the class $\Sigma(A, B, B^*)$ of dissipative systems with collocated actuators and sensors. This class contains many physically motivated examples of strongly stabilizable systems. This second theme is directly linked to the notion of passivity. Systems in the $\Sigma(A, B, B^*)$ class are passive, and this property is exploited frequently to obtain useful results.

2.1 Strong Stability

In this book, we develop system and control theoretic methods for nonexponentially stabilizable systems. In the system theoretic literature on infinite-dimensional linear systems, the prevailing concept has always been exponential stability. In the literature on partial differential equations, however, it is customary to consider merely global asymptotic stability of the trajectories. As a second step, one might try to obtain a uniform rate of decay for the trajectories (for this reason, exponential stability is often called uniform stability in these papers). For many partial differential equations (for instance, undamped wave equations and beam equations with certain boundary conditions), exponential stability is not obtained, even though global asymptotic stability is. The above considerations are concerned only with stability of the state. As we explained in Chapter 1, a system with an exponentially stable semigroup also satisfies certain external stability properties, and these properties are important to arrive at useful system theoretic results. Unfortunately, systems that exhibit only asymptotic convergence of the state to zero do not satisfy these external stability properties, in general. Therefore, an appropriate stability definition should include the behavior of inputs and outputs. To this end, we will introduce the notion of a strongly stable system in Definition 2.1.1 and related stabilizability notions in Definition 2.1.7. A discussion of a number of aspects of the stability definition completes this section. An extensive treatment of the suitability of our definitions of strong stabilizability and strong detectability follows in Chapter 3.

Our definitions apply to linear time-invariant control systems of the form

$$\begin{aligned}
\dot{z}(t) &= Az(t) + Bu(t), \quad z(0) = z_0, \\
y(t) &= Cz(t) + Du(t),
\end{aligned} \tag{2.1}$$

where $z(t) \in Z$, $u(t) \in U$, $y(t) \in Y$ and U, Y, Z are separable Hilbert spaces. Furthermore, A generates a C_0-semigroup of bounded operators $T(t)$, $B \in \mathcal{L}(U, Z)$, $C \in \mathcal{L}(Z, Y)$ and $D \in \mathcal{L}(U, Y)$. We denote this system by $\Sigma(A, B, C, D)$. If the value of D is not of interest, or if $D = 0$, then we denote the system by $\Sigma(A, B, C)$. To define strong stability for such a system we have to introduce a number of maps. First, we define the input map $\Phi_t : \mathbf{L}_2(0, \infty; U) \to Z$ by

$$\Phi_t u = \int_0^t T(t-s) B u(s) ds. \tag{2.2}$$

For all $t \geq 0$, there exists a positive number k_t such that

$$\|\Phi_t u\|_Z \leq k_t \|u\|_2.$$

Next, we define $\tilde{\Phi}_t : \mathbf{L}_2(0, \infty; U) \to Z$ by

$$\tilde{\Phi}_t u = \int_0^t T(s) B u(s) ds. \tag{2.3}$$

Defining the reflection operator $R_t : \mathbf{L}_2(0, \infty; U) \to \mathbf{L}_2(0, \infty; U)$ by

$$(R_t u)(s) := \begin{cases} u(t-s), & s \in [0, t), \\ 0, & s \geq t, \end{cases}$$

we obtain $\|R_t u\|_{L_2(0,t;U)} = \|u\|_{L_2(0,t;U)}$ and $\tilde{\Phi}_t = \Phi_t R_t$. Obviously, this implies that $\|\tilde{\Phi}_t\| = \|\Phi_t\| < k_t$. Under the assumption that the operators $\tilde{\Phi}_t$ are uniformly bounded, i.e., there exists a $k \geq 0$ such that for all u

$$\|\tilde{\Phi}_t u\|_Z \leq k \|u\|_2,$$

we can take the limit for $t \to \infty$ of $\tilde{\Phi}_t$:

$$\tilde{\Phi} u := \lim_{t \to \infty} \tilde{\Phi}_t u = \lim_{t \to \infty} \int_0^t T(s) B u(s) ds \tag{2.4}$$

and $\tilde{\Phi} \in \mathcal{L}(\mathbf{L}_2(0, \infty; U), Z)$. This limiting operator $\tilde{\Phi}$ is called the *extended input map*.

Second, let us introduce the output map $\Psi_t : Z \to \mathbf{L}_2(0, t; Y)$:

$$(\Psi_t z)(\tau) = C T(\tau) z, \quad \tau \in [0, t). \tag{2.5}$$

Again, these operators are bounded and they can be extended to an operator taking values in $\mathbf{L}_2^{loc}(0, \infty; Y)$:

$$\Psi z = \lim_{t \to \infty} \Psi_t z. \tag{2.6}$$

Ψ is called the *extended output map*. If the operators Ψ_t are uniformly bounded, then $\Psi \in \mathcal{L}(Z, \mathbf{L}_2(0, \infty; Y))$.

Finally, we define the input-output map $\mathbb{F}_t : \mathbf{L}_2(0, \infty; U) \to \mathbf{L}_2(0, t; Y)$ by

$$(\mathbb{F}_t u)(\tau) = \int_0^\tau C T(\tau - s) B u(s) ds + D u(\tau), \quad \tau \in [0, t). \tag{2.7}$$

If the operators $\mathbb{F}_t$ are uniformly bounded, then

$$\mathbb{F}u = \lim_{t \to \infty} \mathbb{F}_t u \qquad (2.8)$$

exists, where $\mathbb{F} \in \mathcal{L}(\mathbf{L}_2(0,\infty;U), \mathbf{L}_2(0,\infty;Y))$, and is called the *extended input-output map.*

The operators above stem from the definition of *well-posed linear systems.* This very general class of linear infinite-dimensional systems, which allows for very unbounded input and output operators, was introduced by Weiss (see [106, 107, 108] or the survey by Curtain [24]).

In the case of an exponentially stable semigroup $T(t)$, the maps $\tilde{\Phi}_t$, Ψ_t and $\mathbb{F}_t$ are always uniformly bounded, and so the extended output map Ψ, the extended input map $\tilde{\Phi}$ and the extended input-output map $\mathbb{F}$ all exist and are bounded operators. If $T(t)$ is only strongly stable, then in general this is not the case. This fact motivates the following definition of a strongly stable system.

Definition 2.1.1 *Consider the system* $\Sigma(A, B, C, D)$ *and let* $T(t)$*,* $\tilde{\Phi}$*,* Ψ *and* $\mathbb{F}$ *be defined as above.*

- $T(t)$ *is a* strongly stable semigroup *if for all* $z \in Z$*,* $T(t)z \to 0$ *as* $t \to \infty$.

- $\Sigma(A, B, C, D)$ *is* input stable *if* $\tilde{\Phi} \in \mathcal{L}(\mathbf{L}_2(0,\infty;U), Z)$.

- $\Sigma(A, B, C, D)$ *is* output stable *if* $\Psi \in \mathcal{L}(Z, \mathbf{L}_2(0,\infty;Y))$.

- $\Sigma(A, B, C, D)$ *is* input-output stable *if* $\mathbb{F} \in \mathcal{L}(\mathbf{L}_2(0,\infty;U), \mathbf{L}_2(0,\infty;Y))$.

- $\Sigma(A, B, C, D)$ *is a* strongly stable system *if it is input stable, output stable and input-output stable and if* A *generates a strongly stable semigroup.*

This definition of a strongly stable system was introduced by Staffans in the much more general context of well-posed linear systems, where the input and output operators are allowed to be very unbounded (see, for instance, [97, 98]).

The notions of input stability, output stability and input-output stability can be equivalently expressed in frequency-domain terms. For input stability and output stability this is a direct application of the Paley–Wiener theorem. A proof for the input-output stability part can be found in Weiss [109].

Lemma 2.1.2

- $\Sigma(A, B, C, D)$ *is input stable if and only if* $B^*(sI - A^*)^{-1}z \in \mathbf{H}_2(U)$ *for all* $z \in Z$.

- $\Sigma(A, B, C, D)$ *is output stable if and only if* $C(sI - A)^{-1}z \in \mathbf{H}_2(Y)$ *for all* $z \in Z$.

- $\Sigma(A, B, C, D)$ *is input-output stable if and only if its transfer function* $G(s) = D + C(sI - A)^{-1}B \in \mathbf{H}_\infty(\mathcal{L}(U,Y))$.

It can be easily seen that if A generates an exponentially stable semigroup $T(t)$, $\Sigma(A, B, C, D)$ is input stable, output stable and input-output stable. Hence, every exponentially stable system is a strongly stable system.

The intuition behind the definition of strong stability is that a system is strongly stable if, with any initial condition and any input in $\mathbf{L}_2(0,\infty;U)$, the state is a bounded function of time and the output is in $\mathbf{L}_2(0,\infty;Y)$ and, furthermore, that the state converges to zero whenever the input function equals zero. In fact, it can also be proved that for nonzero

$u \in \mathbf{L}_2(0, \infty; U)$, the state converges to zero. This is done in the following lemma, which will turn out to be of crucial importance when obtaining system and control theoretic results for strongly stable systems.

Lemma 2.1.3 *Assume that the system $\Sigma(A, B, C, D)$ is input stable and that A generates a strongly stable semigroup. Then, for any $u \in \mathbf{L}_2(0, \infty; U)$, the solution $z(t)$ of (2.1) converges to zero as $t \to \infty$.*

Proof The solution $z(t)$ is given by

$$z(t) = T(t)z_0 + \int_0^t T(t-s)Bu(s)ds.$$

The first term tends to zero because $T(t)$ is a strongly stable semigroup. We rewrite the second term in the following way:

$$\int_0^t T(t-s)Bu(s)ds$$
$$= T(t-t_1)\int_0^{t_1} T(t_1-s)Bu(s)ds + \int_{t_1}^t T(t-s)Bu(s)ds.$$

Now, since $T(t)$ is strongly stable and $u(t) \in \mathbf{L}_2(0, \infty; U)$, we have for fixed t_1,

$$\lim_{t \to \infty} T(t-t_1)\int_0^{t_1} T(t_1-s)Bu(s)ds = 0.$$

For the remaining term, we have

$$\int_{t_1}^t T(t-s)Bu(s)ds = \int_0^t T(t-s)Bu_{t_1}(s)ds = (\Phi_t u_{t_1})(t),$$

where we define

$$u_{t_1}(t) = \begin{cases} 0, & t \leq t_1, \\ u(t), & t > t_1 \end{cases}$$

and Φ_t and $\tilde{\Phi}_t$ are defined as in (2.2), (2.3). It is easy to see that $\|\Phi_t\| = \|\tilde{\Phi}_t\| \leq \|\tilde{\Phi}\|$. Hence,

$$\|(\Phi_t u_{t_1})(t)\|_Z \leq \|\tilde{\Phi}\| \left(\int_0^\infty \|u_{t_1}(s)\|^2 dt\right)^{\frac{1}{2}}.$$

By the input stability, $\|\tilde{\Phi}\| < \infty$, so the above expression can be made arbitrarily small by choosing t_1 large enough, which completes our proof. □

The notions of input stability and output stability were also used in Hansen and Weiss [53], but the terminology used there is different. If the system is input stable, then Hansen and Weiss call B an *infinite-time admissible control operator for $T(t)$*. Similarly, our notion of output stability is equivalent to C being an *infinite-time admissible observation operator for $T(t)$*. In the same paper, Hansen and Weiss show that there is a strong relation between input stability and the solvability of a particular Lyapunov equation.

Lemma 2.1.4 (Hansen and Weiss [53, Theorem 3.1]) *System $\Sigma(A, B, C, D)$ is input stable if and only if there exists at least one self-adjoint, nonnegative solution $\Pi \in \mathcal{L}(Z)$ to the Lyapunov equation*

$$A\Pi + \Pi A^* = -BB^*. \tag{2.9}$$

Moreover, in this case, the operator $P \in \mathcal{L}(Z)$ defined by

$$Pz = \tilde{\Phi}\tilde{\Phi}^* z = \int_0^\infty T(t)BB^*T^*(t)z\,dt \qquad (2.10)$$

is the smallest nonnegative solution of (2.9).

By duality, similar statements hold relating output stability to the solvability of the dual Lyapunov equation $A^*Q + QA = -C^*C$.

In addition to the result in Lemma 2.1.4, Hansen and Weiss showed that input stability of $\Sigma(A, B, C, D)$ has a number of implications for the system, in particular for the semigroups $T(t)$ and $T^*(t)$.

Lemma 2.1.5 *Assume that $\Sigma(A, B, C, D)$ is input stable and let P be given by (2.10).*

1. *For all $z \in Z$, $\lim_{t\to\infty} P^{\frac{1}{2}}T^*(t)z = 0$.*

2. *If P is boundedly invertible, then $T^*(t)$ is a strongly stable semigroup.*

3. *If $P > 0$ and $T(t)$ is uniformly bounded, then $T^*(t)$ (and hence also $T(t)$) is a weakly stable semigroup, i.e., $\lim_{t\to\infty}\langle T^*(t)z_1, z_2\rangle = 0$ for all $z_1, z_2 \in Z$.*

4. *If $T^*(t)$ is a strongly stable semigroup, then P is the unique self-adjoint solution of (2.9).*

Again, dual statements hold for output stable systems.

Remark 2.1.6 Note that for an input stable system, approximate controllability is equivalent to $\ker(\tilde{\Phi}^*) = \{0\}$ (see Theorem 4.1.22 in Curtain and Zwart [36] for this equivalence). Therefore, in this case, $\Sigma(A, B, C)$ is approximately controllable if and only if $P > 0$.

In view of the third result in the above lemma, it is interesting to mention the following result (see Remark 3.3 in Hansen and Weiss): Let A be the generator of a weakly stable semigroup $T(t)$. Both $T(t)$ and $T^*(t)$ are *strongly* stable semigroups if

- A has compact resolvent (i.e., for some (and hence for any) $s \in \rho(A)$, the resolvent operator $(sI - A)^{-1}$ is compact)

or

- $\sigma(A) \cap j\mathbb{R}$ is at most countable.

The last condition is a consequence of an important result by Arendt and Batty [2]. They show that if $T(t)$ is uniformly bounded, $\sigma(A) \cap j\mathbb{R}$ is countable and A has no eigenvalues on the imaginary axis, then $T(t)$ is a strongly stable semigroup. Note that their result is applicable, because a weakly stable semigroup is uniformly bounded and the generator of a weakly stable semigroup cannot have purely imaginary eigenvalues.

The result by Arendt and Batty is one instance of a wealth of results about the relation between strong stability of a semigroup and the spectrum of its generator. Unlike the case of exponentially stable semigroups, a necessary and sufficient condition on the spectrum of the infinitesimal generator A for strong stability of the semigroup $T(t)$ generated by A is not known. There are, however, a number of necessary conditions and a number of sufficient conditions known. Huang [54], for instance, has shown that if $T(t)$ is uniformly bounded for $t \geq 0$ and $\sigma(A)$ is contained in the *open* left half-plane, then $T(t)$ is strongly stable. In the same paper he also proved a necessary condition: If $T(t)$ is strongly stable, then it is uniformly bounded and $\sigma(A)$ is contained in the *closed* left half-plane. Batty [15] showed that, in this case, there cannot be eigenvalues on the imaginary axis. An important special case of the result by Arendt and Batty, in view of the class of systems in Section 2.2, is the following: If A generates a contraction semigroup, has compact resolvent and has no eigenvalues on the imaginary axis, then it generates a strongly stable semigroup.

Now that we have a definition of strongly stable systems, we also need related notions of stabilizability and detectability. These are introduced next. In Chapter 3, we will elaborate on why these notions of stabilizability and detectability are appropriate.

Definition 2.1.7

- *The system $\Sigma(A,B,C,D)$ is* strongly stabilizable *if there exists an $F \in \mathcal{L}(Z,U)$ such that $\Sigma(A+BF, B, \begin{bmatrix} F^* & C^* \end{bmatrix}^*, D)$ is output stable and $A+BF$ generates a strongly stable semigroup.*

- *The system $\Sigma(A,B,C,D)$ is* strongly detectable *if there exists an $L \in \mathcal{L}(Y,Z)$ such that $\Sigma(A+LC, \begin{bmatrix} L & B \end{bmatrix}, C, D)$ is input stable and $A+LC$ generates a strongly stable semigroup.*

- *The system $\Sigma(A,B,C,D)$ is* statically stabilizable *if there exists a $K \in \mathcal{L}(Y,U)$ such that $\Sigma(A+BKC, B, C, D)$ is a strongly stable system.*

In [101], Staffans also introduced concepts of strong stabilizability and strong detectability. Even though our definition of a strongly stable system is the same as the one by Staffans [97], our definitions of strong stabilizability and strong detectability are quite different from his, and in Chapter 3 we will discuss the difference. Our definition of static stabilizability is new. It was introduced in Oostveen and Curtain [80], where it was used fruitfully to construct a Riccati theory for strongly stabilizable systems. Moreover, in the following section we identify a class of models that are statically stabilizable.

Our definitions of strongly stable systems (Definition 2.1.1) and of stabilizability and detectability (Definition 2.1.7) are independent of the feedthrough operator D of the system. This is also the case for the definition of static stabilizability. Therefore, it corresponds only to the stabilization by the static output feedback law $u = Ky$ in the special case that $D = 0$. If $D \neq 0$, then the static output feedback $u = (I+KD)^{-1}Ky$ would lead to the strongly stable closed-loop system $\Sigma(A+BKC, B, (I+DK)C, D)$.

In the 1970s and 1980s there has been considerable interest in weak and strong stabilizability of semigroups. Most of these results concern contraction semigroups, or semigroups which are in some sense similar to a contraction. A number of different approaches were used to obtain these results. The earliest papers were written by Slemrod [92, 93]. The paper [92] concerns weak stabilizability of contraction semigroups. In [93], Slemrod used a Lyapunov approach to show that if A is skew adjoint and generates a contraction semigroup, (A, B) is approximately controllable and the trajectories satisfy a certain compactness condition, then the feedback $u = -B^*z$ renders the closed-loop semigroup strongly stable. Another approach was based on a decomposition of the state space in the subspace on which the contraction semigroup is unitary and its orthogonal complement, called the completely nonunitary part of the state space. Many stabilizability results were based on this decomposition, for instance, Benchimol [18, 19], Levan [62, 67] and Levan and Rigby [68]. Benchimol [19] derived a necessary and sufficient condition for strong stabilizability. He showed that if A generates a contraction semigroup and has compact resolvent, then the semigroup generated by $A - BB^*$ is strongly stable if and only if the "weakly unstable states" of $T(t)$ are approximately controllable. The relation between the decomposition approach and the Lyapunov-based approach was studied in Levan [63]. Later, a Riccati equation approach was used to prove the stability of the semigroups generated by $A - BB^*$ and $A - BB^*P$, where P is a nonnegative solution of the standard LQ Riccati equation. Among these papers are Balakrishnan [4] and Levan [64, 65, 66]. A spectral approach was taken by Batty and Phóng [16]. They showed that if A generates a contraction semigroup and $\sigma(A) \cap j\mathbb{R}$ is countable, then the semigroup generated by $A - BB^*$ is strongly stable if and only if there do not exist purely imaginary eigenvalues of A with a corresponding eigenvector in $\ker(B^*)$. The special class of dissipative systems with collocated actuators and sensors, which will be introduced in the next section, consists of systems whose semigroups satisfy the assumptions about strong stabilization of semigroups of most of the above papers, in particular those in Benchimol [19].

In this book, we are interested in the stronger form of stabilizability of the whole system, as in Definition 2.1.7. New results about these notions of stabilizability and detectability are presented in Chapter 3.

It is well known that if $\Sigma(A, B, C)$ is exponentially stabilizable and B is a bounded, finite-rank operator, then A can have only finitely many eigenvalues in the closed right half-plane (see, for instance, Theorem 5.2.3 in Curtain and Zwart [36]). We would like to have similar information for the case of a strongly stabilizable system $\Sigma(A, B, C, D)$ with bounded, finite-rank B. We can obtain this by applying the theorem about the exponentially stabilizable case. This theorem states that if the system $\Sigma(A, B, C, D)$ is β-exponentially stabilizable, then there exists a constant $\delta < \beta$ such that $\Sigma(A) \cap \{s \in \mathbb{C} \mid \mathrm{Re}(s) > \delta\}$ consists of a finite number of eigenvalues with finite multiplicities. A system is called β-exponentially stabilizable if there exists $F \in \mathcal{L}(U, Y)$ such that $e^{\alpha t} T_{BF}(t)$ is exponentially stable for some $\alpha > -\beta$, where $T_{BF}(t)$ is the semigroup generated by $A + BF$. We can apply this theorem with arbitrary $\beta > 0$, because if $T_{BF}(t)$ is a strongly stable semigroup, then $e^{\alpha t} T_{BF}(t)$ is exponentially stable for all $\alpha < 0$. So, according to the theorem, for all $\beta > 0$ there does exist a $\delta < \beta$ such that there are only finitely many points with real part greater than δ in the spectrum of A (counting multiplicities) and all of these spectral points are eigenvalues. Consequently, A has only countably many eigenvalues in the open right half-plane, and in that case these eigenvalues have an accumulation point, which must lie on the imaginary axis. Moreover, there is no continuous or residual spectrum of A in the open right half-plane. If the right half-plane eigenvalues of A do not have an accumulation point on the imaginary axis, then there are only finitely many of them.

2.2   Dissipative Systems with Collocated Actuators and Sensors

By far the most important class of systems that are strongly stabilizable and detectable is the class of dissipative systems with collocated actuators and sensors, which we will often denote as the $\Sigma(A, B, B^*)$ class. Although mathematically very special, they have a prominent place in the engineering literature, mainly as models for flexible structures. The most important mathematical characteristics are that A generates a contraction semigroup and $C = B^*$. The duality condition $C = B^*$ can arise if actuators and sensors are implemented at the same location; hence the terminology "collocation." The term dissipative comes from the fact that the generator of a contraction semigroup is a dissipative operator. These systems are often found in the engineering literature as models in acoustics and for flexible structures (see, for instance, the monograph by Joshi [57], which deals with (finite-dimensional) models of flexible structures). Some simple physical examples of dissipative systems with collocated actuators and sensors are the models for transverse vibrations of a flexible beam with boundary control in Slemrod [94], for vibrations in a flexible plate in You [120] and for the propagation of sound in a tube terminated by linear oscillators in Ito and Propst [56]. At the end of this section, we give the model from Slemrod. The other two models are given in Chapter 9.

An important motivation for studying systems in this class was the benchmark problem SCOLE (Spacecraft Control Laboratory Experiment), which was posed by NASA for the control of flexible spacecraft. This model, which was published in Balakrishnan [5], describes an antenna on a flexible mast which is supported by the space shuttle. This benchmark problem has generated much research in control of flexible structures described by partial differential equations.

The models in the $\Sigma(A, B, B^*)$ class are the classical example of systems that are not exponentially stabilizable. The spectrum of A can have infinitely many points on the imaginary axis or arbitrarily close to it. A case in which this happens is, for instance, when A is skew-adjoint: $\langle Az, z \rangle + \langle z, Az \rangle = 0$ for all $z \in \mathcal{D}(A)$. In this case, all eigenvalues of A are on the imaginary axis. It follows from Theorem 5.2.3 in Curtain and Zwart [36] that if there are infinitely many points of the spectrum of A arbitrarily close to the imaginary axis, then (A, B) cannot be exponentially stabilizable if B is compact, for instance, if B is bounded and has finite rank. In this book, we only consider systems with a bounded input operator B. Note that this does not exclude examples of boundary control. In fact, the first three examples in Chapter 9 are controlled on the boundary.

Even though these systems are not exponentially stabilizable, they do have many useful properties, most importantly regarding stabilizability and robustness. We will give the precise conditions under which we study these systems and then list some important structural properties, many of which will be used intensively throughout this book.

First we need to introduce three concepts which are important for our class of systems.

Definition 2.2.1 *Let A be a possibly unbounded operator on a Hilbert space Z, with domain of definition $\mathcal{D}(A) \subset Z$. A is a dissipative operator if*

$$\langle Az, z \rangle + \langle z, Az \rangle \leq 0 \text{ for all } z \in \mathcal{D}(A). \tag{2.11}$$

Definition 2.2.2 *A linear system $\Sigma(A, B, C, D)$ is passive if there exists a function $S : Z \to \mathbb{R}^+$ (called the storage function) such that the dissipation inequality*

$$S(z(t)) - S(z(0)) \leq \int_0^t \langle u(s), y(s) \rangle ds \tag{2.12}$$

is satisfied for all inputs $u(t)$ and all initial conditions $z(0) \in Z$.

The concept of a passive linear system is closely related to the notion of a positive real transfer function (see Willems [117, 118] for this relation).

Definition 2.2.3 *Let U and Y be real Hilbert spaces. An $\mathcal{L}(U, Y)$-valued transfer function $G(s)$ is positive real if*

1. *$\overline{G(s)} = G(\bar{s})$;*

2. *$G(s)$ is holomorphic on $\mathbb{C}_0^+$;*

3. *$G(s)^* + G(s) \geq 0$ on $\mathbb{C}_0^+$.*

When the third inequality is replaced by $G(s)^* + G(s) \geq \varepsilon I$ for some $\varepsilon > 0$, then we call
$G(s)$ *strictly positive real.*

We need to explain what we mean by $\overline{G(s)}$, as it is not a priori clear what is meant
by the complex conjugate of an operator between two infinite-dimensional spaces. In the
finite-dimensional case, we usually have $U = \mathbb{R}^m$ and $Y = \mathbb{R}^p$, and because s can be
complex, we have to consider the complexification of Y. The same phenomenon occurs in
the infinite-dimensional case. U and Y are real Hilbert spaces, and if s is complex, then
for $u \in U$, $G(s)u \in Y + jY$. Thus, $G(s)$ can be written as $G(s) = G_r(s) + jG_c(s)$, where
both G_r and G_c take values as operators between the *real* spaces U and Y. Now, clearly
$\overline{G(s)}$ can be defined as $\overline{G(s)} = G_r(s) - jG_c(s)$ and the first assumption in Definition 2.2.3
can be reformulated as $G_r(\bar{s}) = G_r(s)$, $G_c(\bar{s}) = -G_c(s)$.

We consider systems given by

$$\begin{aligned}
\dot{z}(t) &= Az(t) + Bu(t), \quad z(0) = z_0, \\
y(t) &= B^* z(t),
\end{aligned} \tag{2.13}$$

where $z \in Z$, $u \in U$, $y \in U$ and Z, U are real separable Hilbert spaces. Furthermore, we
assume the following:

A1. A generates a C_0-semigroup of contractions $T(t)$ on Z.

A2. $B \in \mathcal{L}(U, Z)$.

A3. $\Sigma(A, -, B^*)$ is approximately observable.

A4. $\Sigma(A, B, -)$ is approximately controllable.

A5. A has compact resolvent.

Let us briefly discuss these assumptions. Often, the Hilbert space Z is equipped with
a norm which corresponds to the total energy of the system. In that case, assumption A1
expresses the fact that the energy does not increase if the input is identically equal to
zero. The Lumer–Phillips Theorem (see Pazy [85, Section 1.4]) states that if A generates a
contraction semigroup $T(t)$, then it is dissipative (see Definition 2.2.1). Assumption A5 is
satisfied for most partial differential equations on compact domains. It is well known that
if a closed operator A has compact resolvent, then its spectrum consists only of eigenvalues
of finite multiplicity (see Section III.6.8 in Kato [59]).

In the literature on the control of flexible structures, one often finds systems of the
$\Sigma(A, B, B^*)$ structure, satisfying assumptions A1–A5. The following beam example has
been taken from Slemrod [94], who attributes it to Bailey and Hubbard [3].

Example 2.2.4 Let us consider a flexible cantilever beam of length L. A Euler–
Bernoulli model is used to model the transverse vibrations $w(x,t)$ of the beam. A mass is
attached to the tip of the beam and a piezoelectric film is bonded to one side of the beam.
When a voltage is applied to the film, it will apply a bending moment to the beam. This
voltage is the control input for the system. We measure the angular velocity of the tip. The
following model is suitable for this system:

$$\frac{\partial^4 w}{\partial x^4}(x,t) + \frac{\partial^2 w}{\partial t^2}(x,t) = 0 \quad \text{for } 0 < x < L, \tag{2.14}$$

with boundary conditions

$$w(0,t) = \frac{\partial w}{\partial x}(0,t) = 0, \tag{2.15}$$

$$\frac{\partial^2 w}{\partial x^2}(L,t) = -\frac{\partial^3 w}{\partial t^2 \partial x}(L,t) + u(t), \tag{2.16}$$

$$\frac{\partial^3 w}{\partial x^3}(L,t) = \frac{\partial^2 w}{\partial t^2}(L,t) \tag{2.17}$$

and measurement

$$y(t) = \frac{\partial^2 w}{\partial t \partial x}(L,t). \tag{2.18}$$

Slemrod [94] has shown that the system (2.14)–(2.18) can be represented by the abstract differential equation on a certain Hilbert space Z,

$$\begin{aligned} \dot{z}(t) &= Az(t) + Bu(t), \\ y(t) &= B^* z(t), \end{aligned} \tag{2.19}$$

with a bounded input operator B. The norm on the Hilbert space Z corresponds to the energy in the system. In the same paper, he showed that assumptions A1–A5 are satisfied for this model. Hence, the model has a realization in the $\Sigma(A, B, B^*)$ class. In Chapter 9 we discuss his state-space representation of this model in more detail.   ◇

In Chapter 9, we present three more examples of partial differential equations that fit into the $\Sigma(A, B, B^*)$ framework. It is remarkable that the $\Sigma(A, B, B^*)$ structure, which occurs so frequently in the case of infinite-dimensional systems, is hardly ever encountered in passive finite-dimensional systems. The reason for this is that in case of finite-dimensional systems, one generally uses the standard inner product of $\mathbb{R}^n$. When modelling infinite-dimensional systems it is very common to use an inner product corresponding to the energy in the system. It is only in this energy-related inner product that the system exhibits its special $\Sigma(A, B, B^*)$ structure. Let us illustrate this remark with an example of a mass-spring-damper system.

Example 2.2.5 We consider a mass-spring-damper system, with an external force $u(t)$ acting on the (unit) mass:

$$\ddot{x}(t) + d\dot{x}(t) + kx(t) = u(t).$$

The velocity of the mass is observed:

$$y(t) = \dot{x}(t).$$

If we choose $z = \text{col}(x, \dot{x})$ as the state vector on the state space $\mathbb{R}^2$ with the usual inner product, then we obtain a state-space representation

$$\begin{aligned} \dot{z} &= Az + Bu, \\ y &= Cz, \end{aligned}$$

where

$$A = \begin{bmatrix} 0 & 1 \\ -k & -d \end{bmatrix}, \quad B = \begin{bmatrix} 0 \\ 1 \end{bmatrix}, \quad C = \begin{bmatrix} 0 & 1 \end{bmatrix}.$$

In this case, we do not have $A + A^* \le 0$. However, we could introduce as the state space $\mathbb{R}^2$ with the different inner product

$$\left\langle \begin{pmatrix} x_1 \\ x_2 \end{pmatrix}, \begin{pmatrix} y_1 \\ y_2 \end{pmatrix} \right\rangle = kx_1y_1 + x_2y_2.$$

In this case,

$$\frac{1}{2} \left\| \begin{pmatrix} x \\ \dot{x} \end{pmatrix} \right\|^2 = \frac{1}{2}kx^2 + \frac{1}{2}\dot{x}^2,$$

i.e., the sum of the potential and kinetic energy in the system. On this state space, we still have $C = B^*$, but now,

$$A^* = \begin{bmatrix} 0 & -1 \\ k & -d \end{bmatrix},$$

which leads to

$$A + A^* = \begin{bmatrix} 0 & 0 \\ 0 & -2d \end{bmatrix} \le 0.$$

Thus, we have found a realization satisfying the $\Sigma(A, B, B^*)$ structure, by carefully choosing the inner product on the state space. $\diamond$

Systems in the $\Sigma(A, B, B^*)$ class satisfy a number of useful properties. The most important ones for this book are proven in the following lemma.

Lemma 2.2.6 *Let Z, U and Y be real separable Hilbert spaces and let $A_B := A - BB^*$.*

P1. *If assumptions A1 and A2 hold, then A_B generates a C_0-semigroup of contractions $T_B(t)$. Under the additional assumptions A4 and A5, this semigroup is strongly stable.*

P2. *If assumptions A1 and A2 hold, then A_B^* generates a C_0-semigroup of contractions $T_B^*(t)$. Under the additional assumptions A3 and A5, this semigroup is strongly stable.*

P3. *Under the assumptions A1 and A2, $\Sigma(A, B, B^*)$ and $\Sigma(A_B, B, B^*)$ are passive systems and have positive real transfer functions.*

P4. *Under the assumptions A1 and A2, $\Sigma(A_B, B, B^*)$ is output stable and $\|B^*(sI - A_B)^{-1}z\|_2^2 \le \frac{1}{2}\|z\|^2$.*

P5. *Under the assumptions A1 and A2, $\Sigma(A_B, B, B^*)$ is input stable and $\|B^*(sI - A_B^*)^{-1}z\|_2^2 \le \frac{1}{2}\|z\|^2$.*

P6. *Under the assumptions A1 and A2, $\Sigma(A_B, B, B^*)$ is input-output stable and $\|B^*(sI - A_B)^{-1}B\|_\infty \le 1$.*

Proof P1. By the Lumer–Phillips Theorem, assumption A1 implies that A is a dissipative operator. Now, for $z \in \mathcal{D}(A)$,

$$\begin{aligned} \langle A_B z, z \rangle + \langle z, A_B z \rangle &= \langle Az, z \rangle + \langle z, Az \rangle - 2\|B^*z\|^2 \\ &\le \langle Az, z \rangle + \langle z, Az \rangle \\ &\le 0. \end{aligned}$$

Therefore, A_B is also dissipative. Since $\|T(t)\| = \|T^*(t)\|$, $T^*(t)$ is also a contraction semigroup. Again applying the Lumer–Phillips Theorem, it follows that A^* is dissipative. It now follows in a similar way to that above that A_B^* is also dissipative. Thus, both A_B and A_B^* are dissipative. Applying Corollary 4.4 from Pazy [85], it follows that $T_B(t)$ is a contraction semigroup. $T_B(t)$ was proven to be a strongly stable semigroup in Corollary 3.1 of Benchimol [19], under assumptions A1, A2, A4 and A5.

P2. The proof of P2 is completely analogous to the proof of P1, this time using the assumptions A1, A2, A3 and A5. Note that if A generates a contraction semigroup, then so does A^*, and if A has compact resolvent, then so has A^*.

P3. Assume that assumptions A1 and A2 hold. Let $z_0 \in \mathcal{D}(A)$ and let u be differentiable almost everywhere on $[0, t_1)$ with derivative $\dot{u}(t) \in \mathbf{L}_1(0, t_1; U)$. Then the function $z(t)$ on $[0, t_1)$ given by $z(t) = T(t)z_0 + \int_0^t T(t-s)Bu(s)ds$ is the strong solution of (2.13). In particular, it is differentiable for almost all $t \in [0, t_1)$. Now, using the dissipativity of A,

$$
\begin{aligned}
\frac{d}{dt}\|z(t)\|^2 &= \langle z(t), \dot{z}(t) \rangle + \langle \dot{z}(t), z(t) \rangle \\
&= \langle z(t), Az(t) + Bu(t) \rangle + \langle Az(t) + Bu(t), z(t) \rangle \\
&\leq \langle z(t), Bu(t) \rangle + \langle Bu(t), z(t) \rangle \\
&= 2\langle u(t), y(t) \rangle.
\end{aligned}
$$

Integrating this inequality from 0 to t_1 and defining the storage function $S(z) = \frac{1}{2}\|z\|^2$, we obtain the dissipation inequality

$$
S(z(t_1)) - S(z_0) \leq \int_0^{t_1} \langle u(t), y(t) \rangle dt.
$$

This inequality is extended in the usual way to the case of arbitrary $z_0 \in Z$ and $u \in \mathbf{L}_2(0, \infty; U)$. Hence, the system $\Sigma(A, B, B^*)$ is passive. To prove the positive real property, note that for all $u \in U$ and for $s \in \mathbb{C}_0^+$,

$$
\begin{aligned}
\langle (B^*(sI - A)^{-1}B + B^*(\bar{s}I - A^*)^{-1}B)u, u \rangle \\
= \langle \{(sI - A) + (\bar{s} - A^*)\}(sI - A)^{-1}Bu, (sI - A)^{-1}Bu \rangle \\
= 2\mathrm{Re}(s)\|(sI - A)^{-1}Bu\|^2 - \langle (A + A^*)(sI - A)^{-1}Bu, (sI - A)^{-1}Bu \rangle \\
\geq 2\mathrm{Re}(s)\|(sI - A)^{-1}Bu\|^2 \\
\geq 0.
\end{aligned}
$$

The first inequality follows from the dissipativity of A. The dissipativity also implies that $\mathbb{C}_0^+$ is contained in the resolvent set of A and so $B^*(sI - A)^{-1}B$ is holomorphic in the open right half-plane. Thus, under the assumptions A1 and A2, $B^*(sI - A)^{-1}B$ is positive real. Because $\Sigma(A_B, B, B^*)$ also satisfies assumptions A1 and A2, it follows that it is passive and has a positive real transfer function as well.

P4. For P4, we need to show that $B^*(sI - A_B)^{-1}z \in \mathbf{H}_2(U)$ for all $z \in Z$, or, equivalently, by the Paley–Wiener Theorem (see, for instance, Theorem A.6.21 in Curtain and Zwart [36]), that

$$
\int_0^\infty \|B^*T_B(t)z\|^2 dt < \infty \quad \text{for all } z \in Z.
$$

Now, by assumption A1, A is dissipative. Hence, for all $z \in \mathcal{D}(A)$,

$$\frac{d}{dt}\|T_B(t)z\|^2 = \langle A_B T_B(t)z, T_B(t)z \rangle + \langle T_B(t)z, A_B T_B(t)z \rangle$$
$$= \langle A T_B(t)z, T_B(t)z \rangle + \langle T_B(t)z, A T_B(t)z \rangle - 2\|B^* T_B(t)z\|^2$$
$$\leq -2\|B^* T_B(t)z\|^2.$$

Integrating this inequality, we obtain

$$2\int_0^t \|B^* T_B(s)z\|^2 ds \leq \|z\|^2 - \|T_B(t)z\|^2 \leq \|z\|^2.$$

Since $\mathcal{D}(A)$ is dense in S, this inequality extends to all $z \in Z$. Taking the limit for $t \to \infty$ then leads to

$$\int_0^\infty \|B^* T_B(s)z\|^2 ds \leq \frac{1}{2}\|z\|^2.$$

P5. P5 is proved analogously to P4, by differentiating $\|T_B^*(t)z\|^2$ instead.

P6. For the input-output stability, we refer to a result from Curtain and van Keulen [33]. It states that if $P(s)$ is positive real, then $(I + P(s))^{-1} \in \mathbf{H}_\infty$. Hence, with $P(s) = B^*(sI - A)^{-1}B$, which is positive real by P3,

$$(I + P(s))^{-1} = I - B^*(sI - A_B)^{-1}B \in \mathbf{H}_\infty(\mathcal{L}(U)).$$

Next, we prove that $\|B^*(sI - A_B)^{-1}B\|_\infty \leq 1$. Denoting $G(s) = B^*(sI - A_B)^{-1}B$, we deduce for $\mathrm{Re}(s) > 0$ that

$$\begin{aligned}
G^*(s)G(s) &+ (I - G^*(s))(I - G(s)) \\
&= 2G^*(s)G(s) + I - G(s) - G^*(s) \\
&= I + B^*(\bar{s}I - A_B^*)^{-1} \cdot \\
&\qquad (2BB^* - \bar{s}I + A - BB^* - sI + A - BB^*)(sI - A_B)^{-1}B \\
&= I + B^*(\bar{s}I - A_B^*)^{-1}(-2\mathrm{Re}(s) + A + A^*)(sI - A_B)^{-1}B \\
&\leq I,
\end{aligned}$$

where the inequality follows from the dissipativity of A. Thus,

$$I - G^*(s)G(s) \geq (I - G(s))^*(I - G(s)) \geq 0.$$

So, $G^*(s)G(s) \leq I$ for all $\mathrm{Re}(s) > 0$, which implies that $\|G\|_\infty \leq 1$. $\square$

Summarizing, the above lemma shows that if A generates a contraction semigroup and B is bounded, then $\Sigma(A_B, B, B^*)$ is input stable, output stable and input-output stable. Under the additional assumption that the pair (A, B) is approximately controllable and A has compact resolvent, $\Sigma(A_B, B, B^*)$ is a strongly stable system. Note that the observability assumption A3 was not necessary for this. It was used only to prove the strong stability of the adjoint semigroup $T_B^*(t)$. In fact, the set of assumptions A1, A2, A4 and A5 can be used to prove the strong stability of *both* semigroups $T_B(t)$ and $T_B^*(t)$, using part 3 of Lemma 2.1.5 and Remark 2.1.6. Similarly, the strong stability of both semigroups can be proved to follow from the set of assumptions A1, A2, A3 and A5.

The fact that, under the assumptions A1, A2, A4 and A5, the statement that $\Sigma(A_B, B, B^*)$ is a strongly stable system can be rephrased by saying that the feedback $u = -y$ is strongly stabilizing for $\Sigma(A, B, B^*)$ (the system is statically stabilizable with $K = -I$). It is easy to show that exactly the same holds for the feedback $u(t) = -ky(t)$ for any $k > 0$. The combination of P3 with P6 is a nice instance of the famous passivity theorem. This classical theorem states that the negative feedback interconnection of a positive real system (in our case $\Sigma(A, B, B^*)$) with a strictly positive real system (here, the constant gain I) is input-output stable. This particular property is the most important reason why collocation of actuators and sensors is used so often in the engineering literature on control of flexible structures. Also, it is known that these collocated systems have good robustness properties. For instance, in Curtain and van Keulen [33] it was shown that infinite-dimensional positive real systems can be robustly stabilized with respect to coprime factor perturbations, with a robustness margin of at least $1/\sqrt{2}$. In fact, a controller which achieves this robustness margin is $u(t) = -y(t)$.

We can also deduce information on the location of zeros for single-input–single-output systems in the $\Sigma(A, B, B^*)$ class. Let us consider $G(s) = d + B^*(sI - A)^{-1}B$ with $d \geq 0$. From the proof of Lemma 2.2.6, it follows that

$$G(s) + G(s)^* \geq 2d + 2\mathrm{Re}(s)\|(sI - A)^{-1}B\|.$$

So, if λ is a zero of G, i.e., $G(\lambda) = 0$, then $\mathrm{Re}(\lambda)\|(\lambda I - A)^{-1}B\| \leq -d$. If $d > 0$, then this implies that $\mathrm{Re}(\lambda) < 0$, i.e., G is minimum phase. For the case $d = 0$, let us take an arbitrary λ with $\mathrm{Re}(\lambda) > 0$. As A is dissipative, $\lambda \in \rho(A)$. So, $(\lambda I - A)^{-1}B = 0$ implies that $B = 0$. This contradicts the controllability assumption A3. Thus, in the case $d = 0$, $G(\lambda) = 0$ can hold only for $\mathrm{Re}(\lambda) \leq 0$. We conclude that, if $d = 0$, then $G(s)$ has only zeros in the closed left half-plane, and for $d > 0$ they are in the open left half-plane.

Chapter 3

Strong Stabilizability and Strong Detectability

In the previous chapter, we have set the scene for the remaining chapters. In particular, we have defined the notion of a strongly stable system in Definition 2.1.1. Of course, after defining stability, one should think about suitable definitions of stabilizability and detectability. There are many different candidate definitions for strong stabilizability and strong detectability and the ones that we gave in Definition 2.1.7 may not be the most intuitive ones. Indeed, other definitions may be suitable in certain circumstances. In this chapter, we will elaborate on the definitions of our choice.

3.1 Definitions and Motivation

In Definition 2.1.7, we introduced the following stabilizability and detectability definitions.

- The system $\Sigma(A, B, C)$ is *strongly stabilizable* if there exists an $F \in \mathcal{L}(Z, U)$ such that $\Sigma(A + BF, B, \begin{bmatrix} F^* & C^* \end{bmatrix}^*)$ is output stable and $A + BF$ generates a strongly stable semigroup.

- The system $\Sigma(A, B, C)$ is *strongly detectable* if there exists an $L \in \mathcal{L}(Y, Z)$ such that $\Sigma(A + LC, \begin{bmatrix} L & B \end{bmatrix}, C)$ is input stable and $A + LC$ generates a strongly stable semigroup.

- The system $\Sigma(A, B, C)$ is *statically stabilizable* if there exists a $K \in \mathcal{L}(U, Y)$ such that $\Sigma(A + BKC, B, C)$ is a strongly stable system.

Note that, for notational convenience, we often write

$$\Sigma(A, B, \begin{bmatrix} F^* & C^* \end{bmatrix}^*) \text{ for } \Sigma\left(A, B, \begin{bmatrix} F \\ C \end{bmatrix}\right).$$

The motivation for our definitions of strong stabilizability and strong detectability is twofold. First, our definitions allow us to extend a number of well-known system theoretic results to the case of strongly stabilizable systems. These results are the equivalence of internal and external stability for systems that are stabilizable and detectable and the existence of a unique stabilizing solution to the LQ Riccati equation for a stabilizable and detectable system. Second, we show that if a system is strongly stabilizable and strongly detectable, then there exists a feedback operator F_0 for which $\Sigma(A + BF_0, B, \begin{bmatrix} F_0^* & C^* \end{bmatrix}^*)$ is a strongly stable system. In the following section, we deal with the equivalence between internal and external stability. In Section 3.3, we prove the existence and uniqueness of a strongly stabilizing solution to the LQ Riccati equation and we show that the existence of an operator F_0, as above, is a direct consequence of the results on Riccati equations.

We remark that, contrary to the case of exponential stability, our notions of strong stabilizability and strong detectability are not completely dual. The missing element for duality is that if $A + LC$ generates a strongly stable semigroup, then it does not automatically follow that $A^* + C^*L^*$ generates a strongly stable semigroup. This lack of duality is directly due to the nature of strong convergence and, thus, it is intrinsic to every sensible definition of strong stabilizability.

Before we go on to the next section, it is interesting to compare our definition with the definition of strong stabilizability in Staffans [98]. Although his definition of a strongly stable system is exactly the same as ours, his definition of strong stabilizability is different. After a little reformulation, his definition of strong stabilizability states that there must exist $F \in \mathcal{L}(Z, U)$ such that $\Sigma(A + BF, B, \begin{bmatrix} C^* & F^* \end{bmatrix}^*, D)$ is a strongly stable system. The difference from our definition is obvious: Staffans requires input-output stability and input stability of the closed-loop system, which we do not demand. As we will show in the remainder of this chapter, for many results we do not need input-output stability and input stability to be included in the definition of strong stabilizability.

3.2 Equivalence of Internal and External Stability

In this section, we show that our definitions of strong stabilizability and strong detectability are appropriate to generalize a well-known finite-dimensional result about the equivalence of internal and external stability to the case of strong stability. Let us first formulate this finite-dimensional result.

Theorem 3.2.1 *Consider the finite-dimensional system $\Sigma(A, B, C)$. A is a stable matrix if and only if (A, B) is stabilizable, (A, C) is detectable and $C(sI - A)^{-1}B \in \mathbf{H}_\infty$.*

The version for the case of a strongly stabilizable, strongly detectable systems reads as follows.

Theorem 3.2.2 *$\Sigma(A, B, C)$ is a strongly stable system if and only if it is strongly stabilizable, strongly detectable and input-output stable.*

Proof The sufficiency is obvious, taking $F = 0$ and $L = 0$. For the necessity, assume that $C(sI - A)^{-1}B \in \mathbf{H}_\infty(\mathcal{L}(U, Y))$ and that $\Sigma(A, B, C)$ is strongly stabilizable and strongly detectable. Hence, there exists an $F \in \mathcal{L}(Z, U)$ such that $C(sI - A_F)^{-1}z \in \mathbf{H}_2(Y)$ and $F(sI - A_F)^{-1}z \in \mathbf{H}_2(U)$, and there exists an $L \in \mathcal{L}(Y, Z)$ such that $B^*(sI - A_L^*)^{-1}z \in \mathbf{H}_2(U)$ and $L^*(sI - A_L^*)^{-1}z \in \mathbf{H}_2(Y)$ for all $z \in Z$. Using the perturbation formulas

$$C(sI - A)^{-1}z = C(sI - A_F)^{-1}z - C(sI - A)^{-1}B \cdot F(sI - A_F)^{-1}z,$$
$$B^*(sI - A^*)^{-1}z = B^*(sI - A_L^*)^{-1}z - B^*(sI - A^*)C^* \cdot L^*(sI - A_L^*)^{-1}z,$$

we obtain that for all $z \in Z$ the left-hand sides of these two equations are in $\mathbf{H}_2(Y)$ and $\mathbf{H}_2(U)$, respectively, which proves that $\Sigma(A, B, C)$ is input stable and output stable. To prove the strong stability of the semigroup $T(t)$ generated by A, we consider the perturbation formula

$$T(t)z = T_L(t)z - \int_0^t T_L(t-s)LCT(s)zds.$$

Since $\Sigma(A,B,C)$ is output stable, $CT(\cdot)z \in \mathbf{L}_2(0,\infty;Y)$ and the strong stability of $T_L(t)$ and the input stability of $\Sigma(A+LC,L,-)$ then prove that $T(t)z \to 0$ as $t \to \infty$, using Lemma 2.1.3. □

We remark that if $\Sigma(A,B,C)$ is exponentially stabilizable and detectable in the sense of Definition 5.2.1 of Curtain and Zwart [36], then it is automatically strongly stabilizable and detectable. In [36], they obtained a nice generalization of Theorem 3.2.1 to the case of infinite-dimensional exponentially stabilizable and detectable systems, but the additional assumption that the dimensions of U and Y are finite was needed. Our Theorem 3.2.2 shows that in the case of infinite-dimensional input and output spaces, one does at least obtain strong stability. An analogous result for exponential stability of a large class of systems with unbounded input and output operators is reported by Rebarber [89].

It is easy to see that static stabilizability implies strong stabilizability (taking $F = KC$) and strong detectability (taking $L = BK$). Therefore, the following corollary is immediate.

Corollary 3.2.3 $\Sigma(A,B,C)$ *is a strongly stable system if and only if it is statically stabilizable and input-output stable.*

Let us further examine the connection between static stabilizability and the duo of strong stabilizability and strong detectability. As we just mentioned, static stabilizability implies strong stabilizability and strong detectability. Lemma 3.2.4 gives a complete characterization of the relation between the two concepts.

Lemma 3.2.4 $\Sigma(A,B,C)$ *is statically stabilizable if and only if it is both strongly stabilizable and strongly detectable and there exists a* $K \in \mathcal{L}(Y,U)$ *such that* $G(I - KG)^{-1} \in \mathbf{H}_\infty(\mathcal{L}(U,Y))$*, where* $G(s) = C(sI - A)^{-1}B$*.*

Proof In this proof, we will use the shorthand notation $A_K = A + BKC$, $A_L = A + LC$ and $A_F = A + BF$. First, note that $G(s)(I - KG(s))^{-1} = C(sI - A_K)^{-1}B$. Indeed,

$$\begin{aligned}
G(s)(I - KG(s))^{-1} &= C(sI - A)^{-1}B(I - KC(sI - A)^{-1}B)^{-1} \\
&= C(sI - A)^{-1}B + C(sI - A)^{-1}BKC(sI - A_K)^{-1}B \\
&= C\{(sI - A)^{-1} + (sI - A)^{-1}BKC(sI - A_K)^{-1}\}B \\
&= C(sI - A_K)^{-1}B.
\end{aligned}$$

Necessity: If $\Sigma(A,B,C)$ is statically stabilizable, i.e., if there exists $K \in \mathcal{L}(Y,U)$ such that $\Sigma(A_K,B,C)$ is a strongly stable system, then it is also strongly stabilizable and strongly detectable, taking $F = KC$ and $L = BK$. Furthermore, $G(I - KG)^{-1}$ is the input-output map of the strongly stable system $\Sigma(A_K,B,C)$ and so is in $\mathbf{H}_\infty(\mathcal{L}(U,Y))$.

Sufficiency: Assume that $\Sigma(A,B,C)$ is strongly stabilizable and strongly detectable and that $G(I - KG)^{-1} \in \mathbf{H}_\infty(\mathcal{L}(U,Y))$ for some $K \in \mathcal{L}(Y,U)$. We show that $\Sigma(A,B,C)$ is statically stabilized by the same K. $G(s)(I - KG(s))^{-1}$ is the input-output map of $\Sigma(A_K,B,C)$ and so this system is input-output stable. By the strong stabilizability, there exists $F \in \mathcal{L}(Z,U)$ such that $\Sigma(A_F,B,\begin{bmatrix} F^* & C^* \end{bmatrix}^*)$ is output stable. Thus, we have

$$C(sI - A_K)^{-1}z = C(sI - A_F)^{-1}z + C(sI - A_K)^{-1}(BKC - BF)(sI - A_F)^{-1}z,$$

which shows that $C(sI - A_K)^{-1}z \in \mathbf{H}_2(Y)$ for all $z \in Z$, i.e., $\Sigma(A_K,B,C)$ is output stable. Similarly, the strong detectability implies that there exists an $L \in \mathcal{L}(Y,Z)$ such that

$\Sigma(A_L, [\, B \;\; L \,], C)$ is input stable. Now,

$$B^*(sI - A_K^*)^{-1}z = B^*(sI - A_L^*)^{-1}z$$
$$+ B^*(sI - A_K^*)^{-1}C^*(K^*B^* - L^*)(sI - A_L^*)^{-1}z$$

shows that $B^*(sI - A_K^*)^{-1}z \in \mathbf{H}_2(U)$ for all $z \in Z$, i.e., $\Sigma(A_K, B, -)$ is input stable.

It remains to show that the C_0-semigroup $T_K(t)$ generated by A_K is strongly stable. The detectability assumption also implies that A_L generates a strongly stable semigroup $T_L(t)$. Consider the perturbation formula

$$T_K(t)z = T_L(t)z + \int_0^t T_L(t-s)(BKC - LC)T_K(s)z\,ds.$$

Because $T_L(t)$ is a strongly stable semigroup, $\Sigma(A_L, [\, B \;\; L \,], C)$ is input stable and $\Sigma(A_K, B, C)$ is output stable, we can apply Lemma 2.1.3 to show that $T_K(t)$ is a strongly stable semigroup. □

3.3   The Standard LQ Riccati Equation

In this section, we prove that our definitions of strong stabilizability and strong detectability allow us to generalize an essential result about the existence of a strongly stabilizing solution to the standard LQ Riccati equation from the finite-dimensional case to the case of strongly stabilizable and detectable infinite-dimensional systems. We consider the standard LQ optimal control problem for a system $\Sigma(A, B, C)$, that is, we want to find an input $\bar{u} \in \mathbf{L}_2(0, \infty; U)$ for $\Sigma(A, B, C)$ such that a given quadratic cost criterion is minimized. Let us first formulate the finite-dimensional result.

Theorem 3.3.1 *Consider the cost criterion*

$$\mathcal{J}(z_0, u) = \int_0^\infty (\|y(t)\|^2 + \|u(t)\|^2)dt, \tag{3.1}$$

where $u(t)$ is the input of the finite-dimensional system $\Sigma(A, B, C)$ and $y(t)$ is its output.

1. *If (A, B) is stabilizable, then there exists a minimizing control $\bar{u} \in \mathbf{L}_2(0, \infty; U)$, which is given by $\bar{u}(t) = -B^*X_0 z(t)$, where X_0 is the minimal self-adjoint, nonnegative solution to the algebraic Riccati equation*

$$A^*X + XA - XBB^*X + C^*C = 0. \tag{3.2}$$

   The corresponding minimal cost is $\mathcal{J}(\bar{u}, z_0) = \langle X_0 z_0, z_0 \rangle$.

2. *If, in addition, (A, C) is detectable, then X_0 is the unique self-adjoint, nonnegative solution of (3.2) and $A - BB^*X_0$ is stable.*

For the strongly stabilizable and strongly detectable case, we have the following result.

Theorem 3.3.2 *Consider the system $\Sigma(A, B, C)$ with cost criterion (3.1).*

1. *If $\Sigma(A, B, C)$ is strongly stabilizable, then there exists a minimizing control $\bar{u} \in \mathbf{L}_2(0, \infty; U)$, which is given by $\bar{u}(t) = -B^*X_0 z(t)$, where X_0 is the minimal self-adjoint, nonnegative solution to the algebraic Riccati equation*

$$A^*Xz + XAz - XBB^*Xz + C^*Cz = 0 \tag{3.3}$$

for $z \in \mathcal{D}(A)$. The corresponding minimal cost is $\mathcal{J}(\bar{u}, z_0) = \langle X_0 z_0, z_0 \rangle$. Moreover, $\Sigma(A - BB^* X_0, B, \left[-(B^* X_0)^* \quad C^* \right]^*)$ (i.e., the corresponding closed-loop system) is output stable and input-output stable.

2. *If, in addition, $\Sigma(A, B, C)$ is strongly detectable, then X_0 is the unique self-adjoint, nonnegative solution of (3.4). Moreover, the corresponding closed-loop system is a strongly stable system.*

Proof 1. Let F be the feedback operator that is strongly stabilizing for $\Sigma(A, B, C)$ and choose $u(t) = Fz(t)$ to obtain

$$\mathcal{J}(Fz, z_0) = \int_0^\infty (\|CT_F(t)z_0\|^2 + \|FT_F(t)z_0\|^2) dt$$
$$\leq c\|z_0\|^2$$

for some $c \geq 0$, since $\Sigma(A + BF, B, \left[F^* \quad C^* \right]^*)$ is output stable. So, the system satisfies the conditions of Theorem 6.2.4 in Curtain and Zwart [36] and there exists a minimal self-adjoint, nonnegative solution X_0 of (3.3) such that the minimal cost is $\mathcal{J}(\bar{u}, z_0) = \langle X_0 z_0, z_0 \rangle$. From this we immediately obtain the following estimates, which we will use in the proof of part 2:

$$\int_0^\infty \|B^* X_0 T_{X_0}(t) z_0\|^2 dt \leq \|X_0\| \|z_0\|^2 \tag{3.4}$$

and

$$\int_0^\infty \|C T_{X_0}(t) z_0\|^2 dt \leq \|X_0\| \|z_0\|^2, \tag{3.5}$$

where $T_{X_0}(t)$ is the semigroup generated by $A_{X_0} = A - BB^* X_0$. These estimates show that $\Sigma(A_{X_0}, B, \left[-(B^* X_0)^* \quad C^* \right]^*)$ is output stable.

Next, we prove that the closed-loop system is input-output stable. We define $M(s) = I - B^* X_0(sI - A_{X_0})^{-1} B$ and $N(s) = C(sI - A_{X_0})^{-1} B$. Note that $I - M$ and N are the two components of the closed-loop transfer function, and so we need to show that M and N are in $\mathbf{H}_\infty$. It follows from the output stability of the closed-loop system that $M(s)$ and $N(s)$ are holomorphic for all $s \in \mathbb{C}_0^+$. Hence, for all $s \in \mathbb{C}_0^+$, $M(s)$ and $N(s)$ are well defined and we can deduce the following identity:

$$\begin{aligned}
M(s)^* &M(s) + N(s)^* N(s) \\
&= (I - B^*(\bar{s}I - A_{X_0}^*)^{-1} X_0 B)(I - B^* X_0(sI - A_{X_0})^{-1} B) \\
&\quad + B^*(\bar{s}I - A_{X_0}^*)^{-1} C^* C(sI - A_{X_0})^{-1} B \\
&= I + B^*(\bar{s}I - A_{X_0}^*)^{-1} \{A^* X_0 + X_0 A - X_0 BB^* X_0 + C^* C\}(sI - A_{X_0})^{-1} B \\
&\quad - 2\mathrm{Re}(s) B^*(\bar{s}I - A_{X_0}^*)^{-1} X_0(sI - A_{X_0})^{-1} B \\
&= I - 2\mathrm{Re}(s) B^*(\bar{s}I - A_{X_0}^*)^{-1} X_0(sI - A_{X_0})^{-1} B \\
&\leq I.
\end{aligned}$$

So, for all $s \in \mathbb{C}_0^+$ and all $u \in U$,

$$\|M(s)u\|^2 \leq \|u\|^2 \quad \text{and} \quad \|N(s)u\|^2 \leq \|u\|^2.$$

Thus, both $M(s)$ and $N(s)$ are holomorphic and uniformly bounded in the open right half-plane, i.e., $M \in \mathbf{H}_\infty(\mathcal{L}(U))$ and $N \in \mathbf{H}_\infty(\mathcal{L}(U, Y))$.

2. We first prove the input stability of the closed-loop system, using the perturbation formula

$$B^*(sI - A_{X_0}^*)^{-1}z = B^*(sI - A_L^*)^{-1}z$$
$$-B^*(sI - A_{X_0}^*)^{-1}(X_0 BB^* + C^* L^*)(sI - A_L^*)^{-1}z.$$

The result follows from the input stability of $\Sigma(A_L, [\, B \;\; L \,], -)$ and the input-output stability of $\Sigma(A_{X_0}, B, [\, (B^* X_0)^* \;\; C^* \,]^*)$.

Next, we show that the semigroup $T_{X_0}(t)$ is strongly stable. To do this, we write

$$\dot{z}(t) = (A - BB^* X_0)z(t)$$
$$= (A + LC)z(t) + (-LC - BB^* X_0)z(t)$$

and so

$$z(t) = T_L(t)z_0 - \int_0^t T_L(t-s)[\, L \;\; B \,]\begin{bmatrix} Cz(s) \\ B^* X_0 z(s) \end{bmatrix} ds.$$

Now, by the strong stability of $T_L(t)$, $T_L(t)z_0 \to 0$ as $t \to \infty$. Furthermore, $u_1(t) = Cz(t) \in \mathbf{L}_2(0, \infty; Y)$ by (3.5) and $u_2(t) = B^* X_0 z(t) \in \mathbf{L}_2(0, \infty; U)$ by (3.4). So, we can apply Lemma 2.1.3 to show that the integral term tends to zero as t tends to infinity. Thus, we obtain that $z(t)$ tends to zero as t goes to infinity, i.e., $T_{X_0}(t)$ is strongly stable.

Finally, the uniqueness of X_0 follows in exactly the same way as in the case of an exponentially stable semigroup (see, for instance, Lemma 6.2.6 and Theorem 6.2.7 in Curtain and Zwart [36]).   □

Remark 3.3.3 Note that in the proof of the above theorem, the strong stability of the semigroup generated by A_F was not used. Hence, the result of part 1 of this theorem is already valid if there exists an $F \in \mathcal{L}(Z, U)$ such that $\Sigma(A + BF, B, [\, F^* \;\; C^* \,]^*)$ is output stable. The strong stability of the semigroup generated by A_L was used only to prove the strong stability of the semigroup $T_{X_0}(t)$ and the uniqueness of the solution X_0.

We also need the dual result of Theorem 3.3.2. Because it is completely analogous we state it without proof.

Theorem 3.3.4

1. *If $\Sigma(A, B, C)$ is strongly detectable, then there exists a minimal self-adjoint, nonnegative solution Y_0 to the algebraic Riccati equation*

$$YA^*z + AYz - YC^*CYz + BB^*z = 0 \qquad (3.6)$$

   for $z \in \mathcal{D}(A^*)$. Moreover, $\Sigma(A - Y_0 CC^*, [\, B \;\; -Y_0 C^* \,], C)$ is input stable and input-output stable.

2. *If, in addition, there exists an F such that $(A + BF)^*$ generates a strongly stable semigroup and $\Sigma(A + BF, B, [\, F^* \;\; C^* \,]^*)$ is output stable, then Y_0 is the unique self-adjoint, nonnegative solution of (3.3), $(A - Y_0 CC^*)^*$ generates a strongly stable semigroup and $\Sigma(A - Y_0 CC^*, [\, B \;\; -Y_0 C^* \,], C)$ is input stable, output stable and input-output stable.*

Combining Theorem 3.3.2 with Theorem 3.3.4, we are immediately given the following result.

Corollary 3.3.5 *Suppose that* $\Sigma(A,B,C)$ *is strongly stabilizable and strongly detectable. Then there exists an* $L_0 \in \mathcal{L}(Y,Z)$ *such that* $\Sigma(A + L_0C, \begin{bmatrix} L_0 & B \end{bmatrix}^*, C)$ *is input stable, output stable and input-output stable and there exists an* $F_0 \in \mathcal{L}(Z,U)$ *such that* $\Sigma(A + BF_0, B, \begin{bmatrix} F_0^* & C^* \end{bmatrix}^*)$ *is a strongly stable system.*

Proof The result follows immediately from Theorems 3.3.2 and 3.3.4, by taking $F_0 = -B^*X_0$ and $L_0 = -Y_0C^*$. □

One would like to conclude that $\Sigma(A + L_0C, \begin{bmatrix} L_0 & B \end{bmatrix}^*, C)$ is also a strongly stable system, but $A + L_0C$ does not generate a strongly stable semigroup, in general. This is exactly the lack of duality that was discussed at the end of Section 3.1. On the other hand, from Theorem 3.3.2 it follows that $(A + L_0C)^*$ does generate a strongly stable semigroup. So, in the special case that A has compact resolvent, this implies the strong stability of the semigroup generated by $A + L_0C$.

Example 3.3.6 As an example of this lack of duality for strong stability, consider the shift-semigroup $T(t)$ on the Hilbert space

$$Z = \{f \in \mathbf{L}_2(0,\infty)| \ f \text{ is differentiable and } f' \in \mathbf{L}_2(0,\infty)\},$$

$$\|f\|_Z = \left(\int_0^\infty f(t)^2 dt \right)^{\frac{1}{2}},$$

defined by

$$(T(t)f)(s) = f(s+t) \quad \text{for } t,s \geq 0.$$

It is well known that $\lim_{t\to\infty} T(t)f = 0$ for all $f \in Z$, i.e., T is a strongly stable semigroup on Z. However, $T^*(t)$ is given by

$$(T^*(t)f)(s) = \begin{cases} 0 & \text{for } 0 \leq s < t, \\ f(s-t) & \text{if } s \geq t. \end{cases}$$

Clearly, for all $f \in Z$ and all $t \geq 0$, $\|T^*(t)f\|_Z = \|f\|_Z$. Thus, it follows that $T^*(t)$ is *not* strongly stable. ◇

Finally, we show what results can still be obtained if we weaken the stabilizability and observability assumptions.

Lemma 3.3.7 *Consider the system* $\Sigma(A,B,C)$ *and assume that there exist* $F \in \mathcal{L}(Z,U)$ *and* $L \in \mathcal{L}(Y,Z)$ *such that*

1. $\Sigma_{BF} = \Sigma\left(A + BF, B, \begin{bmatrix} F \\ C \end{bmatrix}\right)$ *is output stable, and*

2. $\Sigma_{LC} = \Sigma(A + LC, \begin{bmatrix} L & B \end{bmatrix}, C)$ *is input stable;*

then $T_{BF}(t)$, *the semigroup generated by* $A + BF$, *is uniformly bounded if and only if the semigroup generated by* $A + LC$, $T_{LC}(t)$, *is also uniformly bounded. Moreover,* $T_{BF}(t)$ *is a strongly stable semigroup if* $T_{LC}(t)$ *is strongly stable.*

Proof Let Ψ_{BF} denote the extended output map of Σ_{BF} and $\tilde{\Phi}_{LC}$ denote the extended input map of Σ_{LC}. These two maps are bounded because of our assumptions. Now, for any $z \in Z$ and $t \geq 0$,

$$T_{BF}(t)z = T_{LC}(t)z + \int_0^t T_{LC}(t-s) \begin{bmatrix} L & B \end{bmatrix} \begin{bmatrix} -C \\ F \end{bmatrix} T_{BF}(s)zds.$$

By the output stability, we have for all $z \in Z$,

$$\tilde{u}(\cdot) := \begin{bmatrix} -C \\ F \end{bmatrix} T_{BF}(\cdot)z = \Psi_{BF}z \in \mathbf{H}_2(Y \times U)$$

and $\|\tilde{u}\|_2 \leq \|\Psi_{BF}\|\|z\|$. Now,

$$\|T_{BF}(t)z\| \leq \|T_{LC}(t)z\| + \left\| \int_0^t T_{LC}(t-s) \begin{bmatrix} L & B \end{bmatrix} \begin{bmatrix} -C \\ F \end{bmatrix} T_{BF}(s)z ds \right\|$$

$$\leq \|T_{LC}(t)\|\|z\| + \|\tilde{\Phi}_{LC}\|\|\Psi_{BF}\|\|z\|.$$

From this we can obtain that

$$\left| \|T_{BF}(t)\| - \|T_{LC}(t)\| \right| \leq \|\tilde{\Phi}_{LC}\|\|\Psi_{BF}\| < \infty.$$

The assertion of strong stability follows directly from Lemma 2.1.3.   □

This lemma has an important consequence for systems $\Sigma(A, B, B^*)$ which only satisfy assumptions A1 and A2. In that case, Lemma 2.2.6 tells us that $\Sigma(A - BB^*, B, B^*)$ is input stable, output stable and input-output stable and the semigroup $T_B(t)$ generated by $A - BB^*$ is a contraction semigroup. By looking at the proof of part 2 of Theorem 3.3.2, it follows that there exists a minimal, self-adjoint, nonnegative solution X_0 to the Riccati equation

$$A^* Xz + X Az - XBB^* Xz + BB^* z = 0,$$

and that the system $\Sigma(A - BB^* X_0, B, [B \ (B^* X_0)^*]^*)$ is output stable. Applying the lemma above, it follows that the semigroup $T_X(t)$ generated by $A - BB^* X_0$ is uniformly bounded. Moreover, if $T_B(t)$ is a strongly stable semigroup, then $T_X(t)$ is also strongly stable. Thus, we have recovered a major part of the results of Theorem 3.3.2 under very weak assumptions.

In this chapter we have only considered stabilization via static feedback. Of course, stabilization by dynamic compensators is also very important and will be studied extensively in Chapter 6.

Chapter 4

Riccati Equations for Statically Stabilizable Systems

In the previous chapter, we derived a result for the existence of a strongly stabilizing solution of the standard LQ Riccati equation for a strongly stabilizable and strongly detectable system. It is important to have results available for more general Riccati equations. It will turn out that nice results can be obtained, using the concept of static stabilizability. Therefore, in this chapter we present a theory for the existence of strongly stabilizing solutions to much more general Riccati equations for statically stabilizable systems.

4.1 Problem Formulation

In this chapter, we consider the following Riccati equation related to the infinite-dimensional system $\Sigma(A, B, C)$:

$$A^*Xz + XAz - (B^*X + NC)^*R^{-1}(B^*X + NC)z + C^*QCz = 0, \quad z \in \mathcal{D}(A). \quad (4.1)$$

Throughout, we assume that $N \in \mathcal{L}(Y, U)$, $Q = Q^* \in \mathcal{L}(Y)$, $R = R^* \in \mathcal{L}(U)$ and that R is boundedly invertible. Our aim is to derive conditions for the existence of stabilizing solutions to this Riccati equation. We define what we mean by a stabilizing solution in the context of strongly stabilizing systems.

Definition 4.1.1 *The operator $X \in \mathcal{L}(Z)$ is a* strongly stabilizing solution *of the Riccati equation (4.1) if*

1. *X satisfies (4.1) for all $z \in \mathcal{D}(A)$, and*

2. *$\Sigma(A_{F_X}, B, \begin{bmatrix} F_X^* & C^* \end{bmatrix}^*)$ is a strongly stable linear system,*

where $A_{F_X} = A + BF_X$ and $F_X = -R^{-1}(B^*X + NC)$.

The approach we take is a spectral factorization approach. This means that the condition for existence of a strongly stabilizing solution will be related to the existence of a spectral factorization of the Popov function associated with the Riccati equation.

Definition 4.1.2 *The* Popov function *associated with the Riccati equation (4.1) is the function $\Pi : j\mathbb{R} \to \mathcal{L}(U)$ given by*

$$\Pi(j\omega) = R + NG(j\omega) + G(j\omega)^*N^* + G(j\omega)^*QG(j\omega), \quad (4.2)$$

where $G(j\omega) = C(j\omega I - A)^{-1}B.$

We remark that a sufficient condition for the Popov function Π to be well defined is that the extended input-output map of $\Sigma(A, B, C)$, $\mathbb{F}$, is a bounded operator from $\mathbf{L}_2(0, \infty; U)$ to $\mathbf{L}_2(0, \infty; Y)$. In this case, $G \in \mathbf{H}_\infty(\mathcal{L}(U, Y))$ and has an extension to the imaginary axis in the sense that $\lim_{\sigma \downarrow 0} G(\sigma + j\omega)u$ exists for all $u \in U$ and for almost all $\omega \in \mathbb{R}$ (see Theorem 4.5 in Rosenblum and Rovnyak [90]). Moreover, in this case, $\Pi \in \mathbf{L}_\infty(\mathcal{L}(U))$.

The type of factorization that we consider is the so-called Wiener–Hopf factorization.

Definition 4.1.3 *Consider a function* $\Pi \in \mathbf{L}_\infty(\mathcal{L}(U))$. *A Wiener–Hopf factorization of* Π *is a factorization*

$$\Pi(j\omega) = M^\sim(j\omega)N(j\omega) \quad \text{for almost all } \omega \in \mathbb{R}, \tag{4.3}$$

where M, N, M^{-1}, $N^{-1} \in \mathbf{H}_\infty(\mathcal{L}(U))$.

A sufficient condition for Π to have a Wiener–Hopf factorization is that it satisfies the *coercivity condition*

$$\Pi(j\omega) \geq \varepsilon I \quad \text{for some } \varepsilon > 0, \text{ almost all } \omega \in \mathbb{R}. \tag{4.4}$$

See Theorem 5 in Devinatz and Shinbrot [40] or Rosenblum and Rovnyak [90, Theorem 3.7].

Particular cases of Wiener–Hopf factorizations are

- spectral factorizations, where $N = M$, and

- J-spectral factorizations, where $N = JM$ and $J = \begin{bmatrix} I & 0 \\ 0 & -I \end{bmatrix}$.

There have been many papers concerning the relation between Riccati equations and spectral factorizations. For the case of finite-dimensional systems, we mention Molinari [74] and Kučera [60]. Callier and Winkin [21] and Weiss [111, 112] deal with the case of exponentially stabilizable infinite-dimensional systems.

More recently, there has been a spate of papers using a spectral factorization approach to solutions of Riccati equations for a class of systems much more general than the class considered here, namely, the class of well-posed linear systems (see Staffans [96, 97, 98, 99], Weiss and Weiss [114] and Mikkola [73]). The generality of the class of well-posed linear systems lies in the fact that B and C (and other operators) are allowed to be very unbounded; in our case, they are assumed to be bounded. So, although our class of systems is more restrictive in that B and C are assumed to be bounded, the type of Riccati equation considered is quite general and most importantly we obtain necessary and sufficient conditions for the existence and uniqueness of strongly stabilizing solutions. Our approach follows the thesis of Weiss [111] very closely (see also his paper [112]).

An essential difference between our approach and the development of the Popov approach in Weiss [111, 112] lies in the initial formulation of the problem. In these papers, he took $C = I$ and this forced him to work with exponential stability, instead of strong stability. As in Weiss and Weiss [114], we utilize the extra C-factor to weaken the stability requirements. Let us consider the cost functional associated with (4.1),

$$\mathcal{J}(z_0, u) = \int_0^\infty \left\langle \begin{bmatrix} C^*QC & C^*N^* \\ NC & R \end{bmatrix} \begin{pmatrix} z(t) \\ u(t) \end{pmatrix}, \begin{pmatrix} z(t) \\ u(t) \end{pmatrix} \right\rangle_{Z \times U} dt$$

subject to the constraints

$$\dot{z}(t) = Az(t) + Bu(t), \quad z(0) = z_0,$$
$$y(t) = Cz(t).$$

Using the extended output map Ψ and the extended input-output map of the system $\Sigma(A, B, C)$, we can write the output y as

$$y(\cdot) = (\Psi z_0)(\cdot) + (\mathbb{F}u)(\cdot).$$

As in Weiss [111] and Weiss and Weiss [114], we reformulate the cost as

$$\mathcal{J}(z_0, u) = \left\langle \begin{bmatrix} \Psi^* Q \Psi & \Psi^*(Q\mathbb{F} + N^*) \\ (\mathbb{F}^* Q + N)\Psi & \mathcal{R} \end{bmatrix} \begin{pmatrix} z_0 \\ u \end{pmatrix}, \begin{pmatrix} z_0 \\ u \end{pmatrix} \right\rangle_{Z \times L_2},$$

where

$$\mathcal{R} = R + N\mathbb{F} + \mathbb{F}^* N^* + \mathbb{F}^* Q \mathbb{F}. \tag{4.5}$$

If we assume that $\Sigma(A, B, C)$ is a strongly stable system, then $\Psi \in \mathcal{L}(Z, \mathbf{L}_2(0, \infty; Y))$ and $\mathbb{F} \in \mathcal{L}(\mathbf{L}_2(0, \infty; U), \mathbf{L}_2(0, \infty; Y))$ and this is sufficient to ensure that $\mathcal{J}$ is well defined (that is, bounded) for all inputs $u \in \mathbf{L}_2(0, \infty; U)$. If $C = I$, then to ensure that Ψ and $\mathbb{F}$ are bounded, one needs to assume that $T(t)$ is exponentially stable, as was done in Weiss [111]. The introduction of the extra C-factor weakens this to the assumptions of output stability and input-output stability.

In Section 4.2, we give the background needed to prove our result. This includes some theory of well-posed and regular linear systems and basic results regarding Toeplitz operators and Wiener–Hopf factorizations. In Section 4.3, we prove the existence and uniqueness of a strongly stabilizing solution to the Riccati equation for strongly stable systems. In Section 4.4, we extend our results to the statically stabilizable case. We illustrate the usefulness of these results by applying them to obtain versions of the positive and bounded real lemma and to obtain results about the standard LQ Riccati equation and the existence of J-spectral factorizations. In Section 4.6, we apply the results to systems in the $\Sigma(A, B, B^*)$ class.

4.2 Mathematical Preliminaries

In this section we give definitions and results needed to prove the main results. Toeplitz operators play a central role in our theory.

Definition 4.2.1 *Let $\Pi \in \mathbf{L}_\infty(\mathcal{L}(Z))$. The* Toeplitz operator *with symbol Π is defined by*

$$\mathcal{T}_\Pi : \mathbf{H}_2(Z) \to \mathbf{H}_2(Z), \quad \mathcal{T}_\Pi z = P_+ \Pi z \quad \text{for } z \in \mathbf{H}_2(Z),$$

where P_+ is the orthogonal projection from $\mathbf{L}_2(-j\infty, j\infty; Z)$ onto $\mathbf{H}_2(Z)$. The time-domain Toeplitz operator *with symbol Π is defined by*

$$\mathcal{T}_\Pi^t : \mathbf{L}_2(0, \infty; Z) \to \mathbf{L}_2(0, \infty; Z), \quad \mathcal{T}_\Pi^t z(\cdot) = \mathfrak{F}^{-1} \mathcal{T}_\Pi (\mathfrak{F}z(\cdot))$$

for $z \in \mathbf{L}_2(0, \infty; Z)$, where $\mathfrak{F}$ denotes the Fourier transformation that maps $\mathbf{L}_2(0, \infty; Z)$ onto $\mathbf{H}_2(Z)$.

That $\mathcal{T}_\Pi$ and $\mathcal{T}_\Pi^t$ are bounded operators follows from Theorem A.6.26 in Curtain and Zwart [36].

We state some properties of Toeplitz operators, which we will use in this chapter. The proofs of the following statements can be found in Section 2.1 of Weiss [111]. For a complex function $G(s)$, we define $G^\sim$ by $G^\sim(s) := G(-\bar{s})^*$.

Lemma 4.2.2

1. *If $G^\sim \in \mathbf{H}_\infty(\mathcal{L}(Z))$ or $K \in \mathbf{H}_\infty(\mathcal{L}(Z))$, then $\mathcal{T}_{GK} = \mathcal{T}_G \mathcal{T}_K$.*

2. *If $G \in \mathbf{H}_\infty(\mathcal{L}(Z))$, then $\mathcal{T}_G$ equals the multiplication map induced by G. Consequently, in this case, $\mathcal{T}_G^t = \mathbb{F}_G$, where $\mathbb{F}_G$ is the extended input-output map of a system with transfer function G.*

3. *The adjoint of the Toeplitz operator $\mathcal{T}_G$ is given by $(\mathcal{T}_G)^* = \mathcal{T}_{G^\sim}$.*

The invertibility of the Toeplitz operator which has the Popov function as its symbol is important in the proof of our main result.

Lemma 4.2.3 *Consider the Popov function Π defined by (4.2) and the operator $\mathcal{R}$ defined by (4.5). Then*

1. *$\mathcal{R}$ equals the time-domain Toeplitz operator $\mathcal{T}_\Pi^t$;*

2. *$\mathcal{R}$ is boundedly invertible if and only if $\mathcal{T}_\Pi$ is;*

3. *$\mathcal{T}_\Pi$ is boundedly invertible if Π admits a Wiener–Hopf factorization.*

Proof 1. This is proven in Lemma 4.27 in Weiss [111].
2. This follows from part 1 and the fact that $\mathcal{T}_\Pi$ and $\mathcal{T}_\Pi^t$ are isomorphic.
3. See Theorem 5 of Devinatz and Shinbrot [40].   □

We shall often use the following result to prove that certain functions are in $\mathbf{H}_\infty$. The result is well known in harmonic analysis and a proof for the scalar case can be found in Duren [42]. The proof for the operator-valued case is analogous and we give it here for completeness.

Lemma 4.2.4 *Let $F : \mathbb{C}_0^+ \to \mathcal{L}(U,Y)$, where U and Y are separable Hilbert spaces. Assume that $F(j\omega) \in \mathbf{L}_\infty(\mathcal{L}(U,Y))$. Recall that $F^\sim(s) := F(-\bar{s})^*$.*

1. *If $F(s)u \in \mathbf{H}_2(Y)$ for all $u \in U$, then $F(s) \in \mathbf{H}_\infty(\mathcal{L}(U,Y))$.*

2. *If $F^\sim(-s)y \in \mathbf{H}_2(U)$ for all $y \in Y$, then $F(s) \in \mathbf{H}_\infty(\mathcal{L}(U,Y))$.*

Proof 1. $F(s)$ is holomorphic for $\mathrm{Re}(s) > 0$, because $F(s)u \in \mathbf{H}_2(Y)$ for all $u \in U$. Let us denote $\|F\|_\infty = M$ and $s = \alpha + j\beta$. By Theorem 11.2 in Duren [42], we have the following Poisson representation of $F(s)$ for $\mathrm{Re}(s) = \alpha > 0$:

$$F(\alpha + j\beta)u = \frac{1}{\pi} \int_{-\infty}^\infty \frac{\alpha}{\alpha^2 + (\beta - \omega)^2} F(j\omega) u\, d\omega.$$

So, for every $s \in \mathbb{C}_0^+$,

$$\|F(s)u\| \le M\|u\| \frac{1}{\pi} \int_{-\infty}^\infty \frac{\alpha}{\alpha^2 + (\beta - \omega)^2} d\omega = M\|u\|.$$

Hence, $F(s)$ is bounded and holomorphic on $\mathbb{C}_0^+$, i.e., $F(s) \in \mathbf{H}_\infty(\mathcal{L}(U,Y))$.

2. It can easily be checked that $F(s) \in \mathbf{L}_\infty(\mathcal{L}(U,Y))$ holds if and only if $F^\sim(-s) \in \mathbf{L}_\infty(\mathcal{L}(U,Y))$ and that $F(s) \in \mathbf{H}_\infty(\mathcal{L}(U,Y))$ if and only if $F^\sim(-s) \in \mathbf{H}_\infty(\mathcal{L}(U,Y))$. Using this, part 2 is a direct consequence of part 1.   □

Next, we need some theory about *well-posed linear systems* and *regular linear systems*. Details and further references can be found in Weiss [110] or in the survey paper by Curtain [24]. For any Hilbert space Z, S_τ is the right shift on $\mathbf{L}_2(0, \infty; Z)$, defined by

$$(S_\tau f)(t) = \begin{cases} 0, & 0 \le t < \tau, \\ f(t - \tau), & t \ge \tau. \end{cases}$$

An operator $\mathbb{F}$ on $\mathbf{L}_2(0, \infty; Z)$ is called *shift invariant* if $\mathbb{F}S_\tau = S_\tau \mathbb{F}$. P_τ denotes the projection of $\mathbf{L}_2(0, \infty; Z)$ onto $\mathbf{L}_2(0, \tau; Z)$ by truncation, defined for $f \in \mathbf{L}_2(0, \infty; Z)$ by

$$(P_\tau f)(t) = \begin{cases} f(t), & 0 \le t < \tau, \\ 0, & t \ge \tau. \end{cases}$$

For $f_1, f_2 \in \mathbf{L}_2(0, \infty; Z)$ and $\tau \ge 0$, the *τ-concatenation* of f_1 and f_2, denoted $f_1 \underset{\tau}{\diamond} f_2$, is defined by

$$(f_1 \underset{\tau}{\diamond} f_2)(t) = \begin{cases} f_1(t), & 0 \le t < \tau, \\ f_2(t - \tau), & t \ge \tau. \end{cases}$$

Note that $f_1 \underset{\tau}{\diamond} f_2 = P_\tau f_1 + S_\tau f_2$. We now define well-posed linear systems.

Definition 4.2.5 *A well-posed linear system on the Hilbert spaces U, Z and Y is a quadruple $(\mathbf{T}, \mathbf{\Phi}, \mathbf{\Psi}, \mathbf{F})$, where*

1. *$\mathbf{T} = \{T(t)\}_{t \ge 0}$, where $T(t)$ is a strongly continuous semigroup of bounded linear operators on Z;*

2. *$\mathbf{\Phi} = \{\Phi_t\}_{t \ge 0}$ is a family of bounded linear operators from $\mathbf{L}_2(0, \infty; U)$ to Z such that*

$$\Phi_{\tau + t}(u \underset{\tau}{\diamond} v) = T(t)\Phi_\tau u + \Phi_t v$$

   for any $u, v \in \mathbf{L}_2(0, \infty; U)$ and any $\tau, t \ge 0$;

3. *$\mathbf{\Psi} = \{\Psi_t\}_{t \ge 0}$ is a family of bounded linear operators from Z to $\mathbf{L}_2(0, \infty; Y)$ such that*

$$\Psi_{\tau + t} z = \Psi_\tau z \underset{\tau}{\diamond} \Psi_t T(\tau) z$$

   for any $z \in Z$ and any $\tau, t \ge 0$ and $\Psi_0 = 0$;

4. *$\mathbf{F} = \{\mathbb{F}_t\}_{t \ge 0}$ is a family of bounded linear operators from $\mathbf{L}_2(0, \infty; U)$ to $\mathbf{L}_2(0, \infty; Y)$ such that*

$$\mathbb{F}_{\tau + t}(u \underset{\tau}{\diamond} v) = \mathbb{F}_\tau u \underset{\tau}{\diamond} (\Psi_t \Phi_\tau u + \mathbb{F}_t v)$$

   for any $u, v \in \mathbf{L}_2(0, \infty; U)$ and any $\tau, t \ge 0$ and $\mathbb{F}_0 = 0$.

The four families of operators defined above relate the input function $u \in \mathbf{L}_2(0, t; U)$, the output function $y \in \mathbf{L}_2(0, t; Y)$, the state at time t, $z(t) \in Z$ and the initial state $z_0 \in Z$ of the system on any finite interval $[0, t)$ in the following way:

$$\begin{pmatrix} z(t) \\ P_t y \end{pmatrix} = \begin{bmatrix} T(t) & \Phi_t \\ \Psi_t & \mathbb{F}_t \end{bmatrix} \begin{pmatrix} z_0 \\ P_t u \end{pmatrix}.$$

It is not a coincidence that we used the same notation for the defining maps of well-posed linear systems and the maps in Section 2.1. With every well-posed linear system we can associate a triple of operators (A, B, C), where B and C are (possibly unbounded) operators from U to Z and Z to Y, respectively. These operators are such that A is the generator of $T(t)$, Φ_t satisfies (2.2) and $\Psi_t z$ satisfies (2.5) for $z \in Z_1 \subset Z$. The space Z_1 equals $D(A)$ equipped with the graph norm of A (see Weiss [107]). If C is bounded, then $Z_1 = Z$. A well-posed linear system is called *regular* if the transfer function $G(s)$ has a strong limit as $s \to \infty$ along the real axis. A sufficient condition for a well-posed linear system to be regular is that the input operator B be bounded. A regular linear system is generated by a quadruple (A, B, C, D) of operators such that (2.2) and (2.5) hold and, furthermore,

$$(\mathbb{F}_t u)(\tau) = C_L \int_0^\tau T(\tau - s)Bu(s)ds + Du(\tau),$$

where $Du = \lim_{s \to \infty} G(s)u$ and C_L is an extension of C to a domain which includes $T(t)z$ for every $z \in Z$ and almost every t (C_L is called the Lebesgue extension of C; see Weiss [107]). If C is bounded, then $C_L = C$ and the formula for $\mathbb{F}$ reduces to (2.7). (A, B, C, D) are called the *generating operators* of the system. Now, we are back in the situation that we described in Chapter 2, with the only exception being that B and C can be unbounded operators. Hence, under the assumption of uniform boundedness of Φ_t, Ψ_t and $\mathbb{F}_t$, we can define the extended input map $\tilde{\Phi}$, the extended output map Ψ and the extended input-output map $\mathbb{F}$ for such a system. Note that in the bounded case, the system is always regular and has the transfer function $G(s) = D + C(sI - A)^{-1}B$. In our applications, B and C will always be bounded but we have to be careful about the boundedness of the input and output operators of realizations of spectral factors. If we wish to emphasize that we consider a system with bounded input and output operators, we use the terminology "bounded linear system."

We also need some results about dual systems for bounded linear systems. The *dual system* $\Sigma^d = \Sigma(A^*, C^*, B^*, D^*)$ can be defined in an obvious way with A^* the generator of the C_0-semigroup $T^*(\cdot)$ on Z. It follows directly from the frequency domain criteria for input stability and output stability in Lemma 2.1.2 that $\Sigma(A, B, C)$ is output stable if and only if $\Sigma(A^*, C^*, B^*)$ is input stable. Moreover, for any $y \in \mathbf{L}_2(0, \infty; Y)$ with derivative $\dot{y} \in \mathbf{L}_2(0, \infty; Y)$ there holds

$$\Psi^* y = \lim_{\tau \to \infty} \int_0^\tau T^*(s)C^* y(s)ds \tag{4.6}$$

and

$$(\mathbb{F}^* y)(t) = (B^* \Psi^* S_t^* y)(t) + D^* y(t) \tag{4.7}$$

$$= B^* \lim_{\tau \to \infty} \int_t^\tau T^*(s - t)C^* y(s)ds + D^* y(t) \tag{4.8}$$

for all $t \geq 0$.

For more general duality results for well-posed linear systems see Section 6 of Weiss and Weiss [114]. The following result was proven in Theorem 11.1 of the same paper (see also Lemma 4.10 of Staffans [101]). It is not particularly obvious, even for bounded linear systems; therefore, we include a proof for this special case.

Lemma 4.2.6 *Let $T(\cdot)$ be a C_0-semigroup and let Ψ be an extended output map for $T(\cdot)$.*

1. *If $\bar{\mathbb{F}} \in \mathcal{L}(\mathbf{L}_2(0,\infty;U), \mathbf{L}_2(0,\infty;Y))$ is shift invariant, then $\Psi^{new} = \bar{\mathbb{F}}^*\Psi$ is an extended output map for $T(\cdot)$.*

2. *Let $(\bar{\mathbf{T}}, \bar{\Phi}, \bar{\Psi}, \bar{\mathbf{F}})$ represent a bounded linear system with the generating operators $(\bar{A}, \bar{B}, \bar{C}, \bar{D})$ and suppose that $\Sigma(\bar{A}, \bar{B}, \bar{C}, \bar{D})$ is output stable. Then, $\Psi^{new} = \bar{\mathbb{F}}^*\Psi = C^{new}T(\cdot)$ is an extended output map for $T(\cdot)$, where*

$$C^{new}z_0 = \bar{B}^* V z_0 + \bar{D}^* C z_0$$

and V satisfies

$$Vz = \bar{\Psi}^*\Psi z = \lim_{\tau \to \infty} \int_0^\tau \bar{T}^*(t)\bar{C}^* C T(t)z \, dt.$$

Proof 1. We need to show that Ψ^{new} satisfies part 3 of Definition 4.2.5, which reduces to showing that

$$S_\tau^*\Psi^{new}z_0 = \Psi^{new}T(\tau)z_0 \quad \text{for } z_0 \in Z.$$

This follows from

$$
\begin{aligned}
S_\tau^*\Psi^{new} &= S_\tau^*\bar{\mathbb{F}}^*\Psi \\
&= \bar{\mathbb{F}}^* S_\tau^*\Psi \quad \text{since } \bar{\mathbb{F}} \text{ is shift invariant} \\
&= \bar{\mathbb{F}}^*\Psi T(\tau) \\
&= \Psi^{new}T(\tau).
\end{aligned}
$$

2. From (2.5) it follows that

$$C^{new}z_0 = (\Psi^{new}z_0)(0) = (\bar{\mathbb{F}}^*\Psi z_0)(0).$$

Taking now $z_0 \in \mathcal{D}(A)$, we see that $w = \Psi z_0$ and $\dot{w} = \Psi A z_0$ are both in $\mathbf{L}_2(0,\infty;Y)$ by assumption. Thus (4.6) and (4.8) hold, yielding

$$
\begin{aligned}
C^{new}z_0 &= (\bar{B}^*\bar{\Psi}^* S_t^*\Psi z_0 + \bar{D}^*\Psi z_0)(0) \quad \text{using (4.7)} \\
&= (\bar{B}^*\bar{\Psi}^*\Psi T(t)z_0 + \bar{D}^* C T(t)z_0)(0) \\
&= (\bar{B}^*\bar{\Psi}^*\Psi + \bar{D}^* C)z_0 \quad \text{using (2.5)}.
\end{aligned}
$$

Since all the operators on the right-hand side are bounded, C^{new} is in $\mathcal{L}(Z,Y)$. The given expression for $V = \bar{\Psi}^*\Psi$ follows from (4.6). □

Finally, we conclude this section with another technical lemma that was proven in Theorem 11.3 in Weiss and Weiss [114]. For completeness, we give a proof for the special case of a bounded linear system.

Lemma 4.2.7 *Let $\Sigma(A,B,C)$ be a strongly stable bounded linear system and let the operators N, Q and R be given. Suppose that the corresponding Popov function $\Pi \in \mathbf{L}_\infty(\mathcal{L}(U))$ has a Wiener–Hopf factorization, $\Pi(j\omega) = M^\sim(j\omega)N(j\omega)$. Then $N(\cdot)$ is the transfer function of a well-posed linear system $(\mathbf{T}, \Phi, \Psi_N, \mathbf{F}_N)$, where the C_0-semigroup and the input map are the same as those for $\Sigma(A,B,C)$.*

Proof Let Π have a Wiener–Hopf factorization $\Pi = M^\sim N$. This implies that M is invertible over $\mathbf{H}_\infty(\mathcal{L}(U))$. We denote the extended input-output map corresponding to the transfer function M by $\mathbb{F}_M$. Because $\mathbb{F}_M$ and M are isomorphic via the Fourier transform (see Theorem A.6.27 in Curtain and Zwart [36]), $\mathbb{F}_M^{-1}$ is bounded and shift invariant.

Next, we define

$$\bar{\mathbb{F}} = (Q\mathbb{F} + N^*)\mathbb{F}_M^{-1}. \tag{4.9}$$

Because $\mathbb{F}$ and $\mathbb{F}_M^{-1}$ are shift invariant and bounded and Q and N are bounded, it follows that $\bar{\mathbb{F}}$ is a shift invariant and bounded operator from $\mathbf{L}_2(0, \infty; U)$ to $\mathbf{L}_2(0, \infty; Y)$. So, from Lemma 4.2.6, we can define a new output map Ψ_N for $T(t)$ by

$$(\Psi_N z_0)(t) = (\bar{\mathbb{F}}^*\Psi z_0)(t) = ((\mathbb{F}_M)^{-*}(\mathbb{F}^*Q + N)\Psi z_0)(t). \tag{4.10}$$

Next, we have to show that $\mathbb{F}_N$ satisfies the functional equation in part 4 of Definition 4.2.5:

$$(\mathbb{F}_N(u \underset{\tau}{\diamond} v))(t) = ((\mathbb{F}_N u) \underset{\tau}{\diamond} (\Psi_N \Phi_\tau u + \mathbb{F}_N v))(t).$$

We can split the above equation into two equations, one for $0 \le t < \tau$ and one for $t > \tau$. The first one is trivially satisfied using the causality of $\mathbb{F}_N$; the second one becomes

$$(\mathbb{F}_N(u \underset{\tau}{\diamond} v))(t) = (\Psi_N \Phi_\tau u + \mathbb{F}_N v)(t - \tau).$$

If we apply a shift over a distance τ to this equation, we see that we must check whether $\mathbb{F}_N$ satisfies

$$S_\tau^* \mathbb{F}_N(u \underset{\tau}{\diamond} v) = \Psi_N \Phi_\tau u + \mathbb{F}_N v,$$

where S_τ is the right shift on $\mathbf{L}_2(0, \infty; U)$. Now, since $(\mathbf{T}, \mathbf{\Phi}, \mathbf{\Psi}, \mathbf{F})$ is a well-posed linear system,

$$S_\tau^* \mathbb{F}(u \underset{\tau}{\diamond} v) = \Psi \Phi_\tau u + \mathbb{F} v$$

and so

$$\bar{\mathbb{F}}^* S_\tau^* \mathbb{F}(u \underset{\tau}{\diamond} v) = \bar{\mathbb{F}}^* \Psi \Phi_\tau u + \bar{\mathbb{F}}^* \mathbb{F} v.$$

From the shift invariance of $\bar{\mathbb{F}}$, we obtain

$$S_\tau^* \bar{\mathbb{F}}^* \mathbb{F}(u \underset{\tau}{\diamond} v) = \bar{\mathbb{F}}^* \Psi \Phi_\tau u + \bar{\mathbb{F}}^* \mathbb{F} v. \tag{4.11}$$

Next,

$$\begin{aligned}
\mathcal{T}_\Pi^t &= \mathcal{T}_{M \sim N}^t \\
&= \mathcal{R} \text{ by part 1 of Lemma 4.2.3} \\
&= (\mathbb{F}^*Q + N)\mathbb{F} + \mathbb{F}^*N^* + R \text{ from (4.5)} \\
&= \mathbb{F}_M^* \bar{\mathbb{F}}^* \mathbb{F} + \mathbb{F}^*N^* + R \text{ from (4.9).}
\end{aligned}$$

By Lemma 4.2.2, $\mathcal{T}_\Pi^t = \mathcal{T}_{M \sim}^t \mathcal{T}_N^t = \mathbb{F}_M^* \mathbb{F}_N$, which in combination with the above shows that

$$\mathbb{F}_M^*(\mathbb{F}_N - \bar{\mathbb{F}}^* \mathbb{F}) = \mathbb{F}^*N^* + R,$$

and since $\mathbb{F}_M$ is invertible we obtain

$$\mathbb{F}_N - \bar{\mathbb{F}}^* \mathbb{F} = \mathbb{F}_M^{-*}(\mathbb{F}^*N^* + R).$$

Using a similar argument as that for $\bar{\mathbb{F}}$ one can show that the right-hand side is invariant with respect to the left shift S_τ^* and so the same holds for the left-hand side. Therefore,

$$S_\tau^*(\mathbb{F}_N - \bar{\mathbb{F}}^*\mathbb{F})(u \underset{\tau}{\diamond} v) = (\mathbb{F}_N - \bar{\mathbb{F}}^*\mathbb{F})S_\tau^*(u \underset{\tau}{\diamond} v)$$
$$= (\mathbb{F}_N - \bar{\mathbb{F}}^*\mathbb{F})v$$

and so

$$S_\tau^*\mathbb{F}_N(u \underset{\tau}{\diamond} v) = S_\tau^*\bar{\mathbb{F}}^*\mathbb{F}(u \underset{\tau}{\diamond} v) + (\mathbb{F}_N - \bar{\mathbb{F}}^*\mathbb{F})v$$
$$= \bar{\mathbb{F}}^*\Psi\Phi_\tau u + \bar{\mathbb{F}}^*\mathbb{F}v + \mathbb{F}_N v - \bar{\mathbb{F}}^*\mathbb{F}v \text{ using (4.11)}$$
$$= \bar{\mathbb{F}}^*\Psi\Phi_\tau u + \mathbb{F}_N v$$
$$= \Psi_N\Phi_\tau u + \mathbb{F}_N v \text{ from (4.11).}$$

This shows that $(\mathbf{T}, \mathbf{\Phi}, \mathbf{\Psi}_N, \mathbf{F}_N)$ is a well-posed linear system with the transfer function $N(s)$. □

In the general theory in Weiss and Weiss [114], neither N nor M needs to be regular. However, in the case of a bounded linear system, N is always regular, since it has the same bounded input operator B as the system.

4.3 Riccati Equations for Stable Systems

In this section we will prove our main result of this chapter, which relates strongly stabilizing solutions of the Riccati equation to Wiener–Hopf factorizations. However, before doing so, we prove that the Riccati equation has at most one strongly stabilizing solution. The proof is completely analogous to the one of Lemma 3 in Curtain and Zwart [37].

Lemma 4.3.1 *If the Riccati equation (4.1) has a self-adjoint solution X which is such that $A + BF_X$ generates a strongly stable semigroup, where $F_X = -BR^{-1}(B^*X + NC)$, then it is the only self-adjoint solution with this property.*

Proof Let $X_1 \in \mathcal{L}(Z)$ and $X_2 \in \mathcal{L}(Z)$ be two such solutions of (4.1). Let $F_{X_i} = -BR^{-1}(B^*X_i + NC)$, $i = 1, 2$, and let $T_{F_{X_i}}(t)$ be the strongly stable semigroup generated by $A + BF_i$. The Riccati equation (4.1) can be rewritten in the following ways:

$$(A + BF_{X_1})^*X_1 z + X_1(A + BF_{X_2})z + X_1 BR^{-1}B^*X_2 z + C^*(Q - N^*R^{-1}N)Cz = 0,$$
$$(A + BF_{X_1})^*X_2 z + X_2(A + BF_{X_2})z + X_1 BR^{-1}B^*X_2 z + C^*(Q - N^*R^{-1}N)Cz = 0$$

for all $z \in \mathcal{D}(A)$. Subtracting the two equations above, we obtain

$$(A + BF_{X_1})^*(X_1 - X_2) + (X_1 - X_2)(A + BF_{X_2}) = 0.$$

This implies that for all $z_1, z_2 \in \mathcal{D}(A)$ there holds

$$\frac{d}{dt}\langle T_{F_{X_1}}(t)z_1, (X_1 - X_2)T_{F_{X_2}}(t)z_2 \rangle = 0.$$

Thus, $\langle T_{F_{X_1}}(t)z_1, (X_1 - X_2)T_{F_{X_2}}(t)z_2 \rangle$ is equal to some constant, which must be 0 by the strong stability of $T_{F_{X_1}}(t)$ and $T_{F_{X_2}}(t)$. Substituting $t = 0$ and noting that $\mathcal{D}(A)$ is dense in Z, we obtain that $X_1 = X_2$. □

Next, we prove our main result of this chapter. The sufficiency part is proven in Theorem 16 of Staffans [99] by a different approach. Necessary and sufficient conditions can also be found in the report by Mikkola [73].

Theorem 4.3.2 *Consider the strongly stable system $\Sigma(A, B, C)$ and the operators $N \in \mathcal{L}(Y, U)$, $Q = Q^* \in \mathcal{L}(Y)$, $R = R^* \in \mathcal{L}(U)$, with R boundedly invertible. There exists a unique, self-adjoint, strongly stabilizing solution to the Riccati equation (4.1) if and only if the associated Popov function Π has a Wiener–Hopf factorization.*

Proof *Necessity:* Suppose that the Riccati equation has a strongly stabilizing solution X with corresponding feedback operator F_X. Then, with a long but straightforward calculation, using the fact that X satisfies the Riccati equation, we can check that the following expression holds:

$$\Pi(j\omega) = (I - F_X(j\omega I - A)^{-1}B)^* R(I - F_X(j\omega I - A)^{-1}B).$$

We define $N(s) = I - F_X(sI - A)^{-1}B$, $M(s) = RN(s)$. To show that N and M form a Wiener–Hopf factorization of Π, it is sufficient to show that $N(s) \in \mathbf{H}_\infty(\mathcal{L}(U))$ and $N^{-1}(s) \in \mathbf{H}_\infty(\mathcal{L}(U))$, because the invertibility of R shows that the same then holds for M and M^{-1}.

Because X is a strongly stabilizing solution, the corresponding closed-loop system is input-output stable. As a consequence, $F_X(sI - A_{F_X})^{-1}B \in \mathbf{H}_\infty(\mathcal{L}(U))$ and so $N^{-1}(s) = I + F_X(sI - A_{F_X})^{-1}B \in \mathbf{H}_\infty(\mathcal{L}(U))$. To prove that $N \in \mathbf{H}_\infty(\mathcal{L}(U))$, note that $B^*(sI - A^*)^{-1}z \in \mathbf{H}_2(U)$ implies that $(I - N^\sim(-s))u = B^*(sI - A^*)^{-1}F_X^*u$ is in $\mathbf{H}_2(U)$ for all $u \in U$. Moreover, $N(s) = (M^\sim)^{-1}\Pi \in \mathbf{L}_\infty(\mathcal{L}(U))$ and consequently $N^\sim(-s) \in \mathbf{L}_\infty(\mathcal{L}(U))$, as well. So, by Lemma 4.2.4, $N(s) \in \mathbf{H}_\infty(\mathcal{L}(U))$.

Sufficiency: By assumption, $\Sigma(A, B, C)$ is a strongly stable system. Suppose that Π has a Wiener–Hopf factorization, $\Pi(j\omega) = M^\sim(j\omega)N(j\omega)$. By part 3 of Lemma 4.2.3, $\mathcal{T}_\Pi$ is boundedly invertible and by part 2 of the same lemma it follows that $\mathcal{R}$ is boundedly invertible. As in the proof of Theorem 2.15 in Weiss [111], one can show that the self-adjoint operator X defined by

$$X = \Psi^*Q\Psi - \Psi^*(Q\mathbb{F} + N^*)\mathcal{R}^{-1}(\mathbb{F}^*Q + N)\Psi \tag{4.12}$$

is in $\mathcal{L}(Z)$ and satisfies the Riccati equation (4.1). By the boundedness of all operators involved, it follows that $\bar{u} \in \mathbf{L}_2(0, \infty; U)$ and $\bar{y} \in \mathbf{L}_2(0, \infty; Y)$, where

$$\bar{u}(\cdot) = -R^{-1}(B^*X + NC)\bar{z}(\cdot) = F_X T_{F_X}(\cdot)z_0$$
$$= -\mathcal{R}^{-1}(\mathbb{F}^*Q + N)\Psi z_0, \tag{4.13}$$
$$\bar{y}(\cdot) = C\bar{z}(\cdot) = CT_{F_X}(\cdot)z_0 = \Psi z_0 + \mathbb{F}\bar{u}. \tag{4.14}$$

Because of the C-factor we cannot conclude exponential stability from (4.14), as was done in Weiss [111, 112], but we can show that T_{F_X} is a strongly stable semigroup by considering the perturbation formula from Theorem 3.2.1 in Curtain and Zwart [36]:

$$T_{F_X}(t)z_0 = T(t)z_0 + \int_0^t T(t-s)BF_X T_{F_X}(s)z_0 ds$$
$$= T(t)z_0 + \int_0^t T(t-s)B\bar{u}(s)ds \text{ using (4.13)}.$$

Now, since $\Sigma(A, B, C)$ is a strongly stable system and $\bar{u}(t) \in \mathbf{L}_2(0, \infty; U)$, Lemma 2.1.3 shows that the right-hand side tends to zero as t goes to infinity. Hence, $T_{F_X}(t)$ is a strongly stable semigroup.

Next, we prove the input-output stability of the optimal closed-loop system, which is the system given in part 2 of Definition 4.1.1. That is, we show that $F_X(sI - A_{F_X})^{-1}B$ and $C(sI - A_{F_X})^{-1}B \in \mathbf{H}_\infty$. Let us start with the first one. From Lemma 4.2.7 we have that N has a realization $(\mathbf{T}, \mathbf{\Phi}, \mathbf{\Psi}_N, \mathbf{F}_N)$ as a well-posed linear system. It follows from this that N has A as generator of its semigroup and that B is its input operator. Because B is bounded, N actually has a realization as a regular linear system with generating operators (A, B, C_N, D_N); see Theorem 5.8 in Weiss [111]. If we normalize N such that $\lim_{\lambda \to \infty} N(\lambda)u = u$ (this can be done because N is invertible), then clearly $D_N = I$. To find C_N, note that $\mathcal{R}^{-1}(\mathbb{F}^*Q + N)\Psi z_0 = \mathcal{R}^{-1}(B^*X + NC)T_{F_X}(\cdot)z_0$ and $\mathcal{R} = \mathbb{F}_M^*\mathbb{F}_N$. Thus, using the defining equation (4.10) for Ψ_N, we obtain

$$(\Psi_N z_0)(t) = \mathbb{F}_N(R^{-1}(B^*X + NC)T_{F_X}(\cdot)z_0)(t).$$

Using the regularity of $\mathbb{F}_N$, with $D = I$, and taking $t = 0$ in the above equation we see immediately that

$$C_N z_0 = R^{-1}(B^*X + NC)z_0 = -F_X z_0.$$

So, $N(s) = I - F_X(sI - A)^{-1}B$. Because N is a Wiener–Hopf factor, it follows that $N^{-1}(s) = I + F_X(sI - A_{F_X})^{-1}B \in \mathbf{H}_\infty(\mathcal{L}(U))$. To prove that also $C(sI - A_{F_X})^{-1}B \in \mathbf{H}_\infty$, we use the perturbation formula

$$C(sI - A_{F_X})^{-1}B = C(sI - A)^{-1}B + C(sI - A)^{-1}B \cdot F_X(sI - A_{F_X})^{-1}B.$$

$C(sI - A)^{-1}B \in \mathbf{H}_\infty(\mathcal{L}(U, Y))$ because of the strong stability of $\Sigma(A, B, C)$ and we have just proved that $F_X(sI - A_{F_X})^{-1}B \in \mathbf{H}_\infty(\mathcal{L}(U))$. It now follows from the perturbation formula above that $C(sI - A_{F_X})^{-1}B \in \mathbf{H}_\infty(\mathcal{L}(U, Y))$. Thus, we have proven the input-output stability of the closed-loop system.

Next, we prove that the closed-loop system is input stable. To do so, we use the perturbation formula

$$B^*(sI - A_{F_X}^*)^{-1}z = B^*(sI - A^*)^{-1}z + B^*(sI - A_{F_X}^*)^{-1}F_X^* \cdot B^*(sI - A^*)^{-1}z.$$

Now, the input stability of $\Sigma(A, B, C)$ combined with the input-output stability of the closed-loop system shows that for all $z \in Z$, $B^*(sI - A_{F_X}^*)^{-1}z \in \mathbf{H}_2(U)$.

The output stability of the closed-loop system follows directly from (4.13) and (4.14). The uniqueness of the strongly stabilizing solution is proven in Lemma 4.3.1. □

In the coercive case, we obtain the following corollary.

Corollary 4.3.3 *Consider the strongly stable system $\Sigma(A, B, C)$ and the operators N, Q, R. There exists a unique, self-adjoint, strongly stabilizing solution to the Riccati equation (4.1) if the associated Popov function Π is coercive. Conversely, if there exists a unique, self-adjoint, strongly stabilizing solution to the Riccati equation (4.1) and if $R \geq \mu I$ for some $\mu > 0$ (i.e., R is coercive), then the associated Popov function Π is coercive.*

Proof Assume that Π is coercive. Theorem 3.7 in Rosenblum and Rovnyak [90] shows that Π has a Wiener–Hopf factorization. Theorem 4.3.2 then implies that (4.1) has a unique, self-adjoint, strongly stabilizing solution.

For the converse, assume that R is coercive and that (4.1) has a self-adjoint, strongly stabilizing solution X, with corresponding feedback F_X. In the proof of Theorem 4.3.2 it is shown that

$$\Pi(j\omega) = S^\sim(j\omega)RS(j\omega),$$

where $S(s) = I - F_X(sI - A)^{-1}B$ and $S, S^{-1} \in \mathbf{H}_\infty(\mathcal{L}(U))$. Because R is coercive and S is invertible over $\mathbf{H}_\infty(\mathcal{L}(U))$, Π is also coercive. □

4.4   Riccati Equations for Unstable Systems

Next, we turn our attention to the strongly stabilizing solution for systems that are not stable. The following lemma shows that in certain cases, we can reduce the unstable case to the stable case.

Lemma 4.4.1

1. $X = X^* \in \mathcal{L}(Z)$ satisfies the Riccati equation (4.1) if and only if X satisfies the feedback Riccati equation

$$A_F^* Xz + X A_F z - (B^* X + NC + RF)^* R^{-1} (B^* X + NC + RF)z$$
$$+ (C^* QC + C^* N^* F + F^* NC + F^* RF)z = 0 \qquad (4.15)$$

   for all $z \in \mathcal{D}(A)$, where $A_F = A + BF$.

2. The Popov function Π associated with (4.1) is related to the Popov function Π_F associated with (4.15) by

$$\Pi_F(j\omega) = R + (NC + RF)(j\omega I - A_F)^{-1} B$$
$$+ B^*(-j\omega I - A_F^*)^{-1} (F^* R + C^* N^*) \qquad (4.16)$$
$$+ B^*(-j\omega I - A_F^*)^{-1} (F^* NC + C^* N^* F + F^* RF + C^* QC)$$
$$\cdot (j\omega I - A_F)^{-1} B$$
$$= S_F^{\sim}(j\omega) \Pi(j\omega) S_F(j\omega) \qquad (4.17)$$

   for $j\omega \in \rho(A) \cap \rho(A_F)$, where $S_F(s) = I + F(sI - A_F)^{-1} B$.

Proof Part 1 was established as Lemma 23 in Weiss [112] and part 2 as Lemma 24 in the same paper. His proofs are a matter of algebraic manipulations. The main point is that the feedback operator F_X corresponding to (4.1) is related to the feedback operator F_{X_F} corresponding to (4.15) by $F_X = F + F_{X_F}$. □

To retain the original form of the Riccati equation, we must use feedbacks of the form $F = KC$. In this case, we can extend the result of the previous lemma, as follows.

Lemma 4.4.2 *Let $F = KC$ for some $K \in \mathcal{L}(Y, U)$. X is a strongly stabilizing solution for (4.1) if and only if it is a strongly stabilizing solution for (4.15). Furthermore, the generator of the strongly stable semigroup is the same in both cases.*

Proof Let us denote $A_F = A + BF$ and let the feedback operator corresponding to (4.1) be F_X and the feedback operator corresponding to (4.15) be F_{X_F}. They are related by $F_X = F + F_{X_F}$. This implies that the generators of the closed-loop semigroups are equal: $A_F + B F_{X_F} = A + B F_X$. Hence, we need to show that $\Sigma(A + B F_X, B, [\ C^* \quad F_X^* \ ]^*)$ is a strongly stable system if and only if the system $\Sigma(A + B F_X, B, [\ C^* \quad F_X^* - F^* \ ]^*)$ is strongly stable. This follows directly, because we took $F = KC$. □

The above lemmas allow us to reduce the problem of a Riccati equation for an unstable system to one for a stable system, provided that the unstable system is statically stabilizable. The sufficiency part of the following theorem is proven in Theorem 16 of Staffans [99], using a different approach.

Theorem 4.4.3 *Consider the Riccati equation (4.1), and assume that $\Sigma(A, B, C)$ is statically stabilizable by a static output feedback $u = Ky$. Let Π_{KC} be the transformed Popov function, as in (4.17), with $F = KC$.*

1. *Equation (4.1) has a unique, self-adjoint, strongly stabilizing solution if and only if Π_{KC} has a Wiener–Hopf factorization.*

2. *For the coercive case, define the quadratic form*

$$\mathcal{F}(y,u) = \langle Qy,y \rangle + \langle Ny,u \rangle + \langle u,Ny \rangle + \langle Ru,u \rangle.$$

If there exists an $\varepsilon > 0$ such that for all $(\omega,z,u,y) \in \mathbb{R} \times \mathcal{D}(A) \times U \times Y$ satisfying $j\omega z = Az + Bu$, $y = Cz$, there holds $\mathcal{F}(y,u) \geq \varepsilon(\|y\|^2 + \|u\|^2)$, then the Riccati equation (4.1) has a unique, self-adjoint, strongly stabilizing solution. Conversely, if $R \geq \mu I$ for some $\mu > 0$ and the Riccati equation (4.1) has a unique, strongly stabilizing solution, then there exists an $\varepsilon > 0$ such that $\mathcal{F}(y,u) \geq \varepsilon(\|y\|^2 + \|u\|^2)$ for all $(\omega,z,u,y) \in \mathbb{R} \times \mathcal{D}(A) \times U \times Y$ satisfying $j\omega z = Az + Bu$, $y = Cz$.

Proof 1. Theorem 4.3.2 shows that the feedback Riccati equation (4.15) has a unique, self-adjoint, strongly stabilizing solution if and only if Π_{KC} has a Wiener–Hopf factorization. Lemma 4.3.1, Lemma 4.4.1 and Lemma 4.4.2 show that the feedback Riccati equation (4.15) has a unique, self-adjoint, strongly stabilizing solution if and only if the Riccati equation (4.1) has one.

2. Let $A_{KC} = A + BKC$. We introduce a second quadratic form

$$\mathcal{F}_K(y,v) = \langle Q_K y,y \rangle + \langle N_K y,v \rangle + \langle v,N_K y \rangle + \langle Rv,v \rangle,$$

where $Q_K = Q + K^*N + N^*K + K^*RK$ and $N_K = N + RK$. The proof will consist of the following three steps:

(a) First we show that there exists an $\varepsilon > 0$ such that $\mathcal{F}(y,u) \geq \varepsilon(\|y\|^2 + \|u\|^2)$ for all $(\omega,z,u,y) \in \mathbb{R} \times \mathcal{D}(A) \times U \times Y$ satisfying $j\omega z = Az + Bu$, $y = Cz$ if and only if there exists a $\delta > 0$ such that $\mathcal{F}_K(y,v) \geq \delta(\|y\|^2 + \|v\|^2)$ for all $(\omega,z,v,y) \in \mathbb{R} \times \mathcal{D}(A) \times U \times Y$ satisfying $j\omega z = A_{KC}z + Bv$, $y = Cz$.

(b) Second, we show that there exists a $\delta > 0$ such that $\mathcal{F}_K(y,v) \geq \delta(\|y\|^2 + \|v\|^2)$ for all $(\omega,z,v,y) \in \mathbb{R} \times \mathcal{D}(A) \times U \times Y$ satisfying $j\omega z = A_{KC}z + Bv$, $y = Cz$ if and only if Π_{KC} is coercive.

(c) Finally, we show that there exists a strongly stabilizing solution if Π_{KC} is coercive, and conversely that Π_{KC} is coercive if there exists a strongly stabilizing solution and R is coercive.

(a) Note that if we take $u = Ky + v$, then $j\omega z = Az + Bu$, $y = Cz$ if and only if $j\omega z = A_{KC}z + Bv$, $y = Cz$. Furthermore, in this case, $\mathcal{F}(y,u) = \mathcal{F}_K(y,v)$ and

$$\|u\|^2 = \|Ky + v\|^2 \leq 2\|K\|^2\|y\|^2 + 2\|v\|^2,$$
$$\|v\|^2 = \|-Ky + u\|^2 \leq 2\|K\|^2\|y\|^2 + 2\|u\|^2.$$

So, taking $\alpha = \max(2, 1 + 2\|K\|^2)$, we have

$$\|y\|^2 + \|u\|^2 \leq (1 + 2\|K\|^2)\|y\|^2 + 2\|v\|^2 \leq \alpha(\|y\|^2 + \|v\|^2),$$
$$\|y\|^2 + \|v\|^2 \leq (1 + 2\|K\|^2)\|y\|^2 + 2\|v\|^2 \leq \alpha(\|y\|^2 + \|u\|^2).$$

Assume now that $\mathcal{F}(y,u) \geq \varepsilon(\|y\|^2 + \|u\|^2)$; then

$$\mathcal{F}_K(y,v) = \mathcal{F}(y,Ky + v) \geq \varepsilon(\|y\|^2 + \|Ky + v\|^2) \geq \frac{\varepsilon}{\alpha}(\|y\|^2 + \|v\|^2).$$

Conversely, if $\mathcal{F}_K(y,v) \geq \delta(\|y\|^2 + \|v\|^2)$, then $\mathcal{F}(y,u) \geq \frac{\delta}{\alpha}(\|y\|^2 + \|u\|^2)$.

(b) Let Π_{KC} be coercive, i.e., there exists $\varepsilon > 0$ such that for all $v \in U$ and almost all $\omega \in \mathbb{R}$, $\langle \Pi_{KC}(j\omega)v, v \rangle \geq \varepsilon \|v\|^2$. Let (y, v) be such that there exist ω and z satisfying $j\omega z = A_{KC}z + Bv$, $y = Cz$. It is shown by Huang [54] that since A_{KC} generates a strongly stable semigroup, on the imaginary axis it can have only continuous spectrum. Therefore, for all $\omega \in \mathbb{R}$, $(j\omega I - A_{KC})^{-1}$ exists, though it may be unbounded. Consequently, y can be expressed as $y = C(j\omega I - A_{KC})^{-1} Bv$. From the strong stability of $\Sigma(A_{KC}, B, C)$ it follows that $C(sI - A_{KC})^{-1} B \in \mathbf{H}_\infty(\mathcal{L}(U, Y))$. Hence,

$$\|y\| = \|C(j\omega I - A_{KC})^{-1} Bv\| \leq \|C(j\omega I - A_{KC})^{-1} B\|_\infty \|v\| = c\|v\|.$$

Therefore,

$$(c^2 + 1)\|v\|^2 \geq \|y\|^2 + \|v\|^2.$$

Hence, choosing $\delta = \varepsilon/(c^2 + 1)$,

$$\mathcal{F}_K(y, v) = \mathcal{F}_K(C(j\omega I - A_{KC})^{-1} Bv, v) = \langle \Pi_{KC}(j\omega)v, v \rangle$$
$$\geq \varepsilon\|v\|^2 \geq \delta(\|y\|^2 + \|v\|^2).$$

Conversely, if we assume that $\mathcal{F}_K(y, v) \geq \delta(\|y\|^2 + \|v\|^2)$, then for almost all $\omega \in \mathbb{R}$, we have

$$\langle \Pi_{KC}(j\omega)v, v \rangle = \mathcal{F}_K(C(j\omega - A_{KC})^{-1} Bv, v)$$
$$\geq \delta(\|C(j\omega - A_{KC})^{-1} Bv\|^2 + \|v\|^2) \geq \delta\|v\|^2,$$

i.e., Π_{KC} is coercive.

(c) This part of the proof is a direct consequence of Corollary 4.3.3, Lemma 4.4.1 and Lemma 4.4.2. □

4.5   Special Types of Riccati Equations

In this section we show how the general results of Sections 4.3 and 4.4 apply to certain special types of Riccati equations. First we use the results of Section 4.3 to derive a Riccati criterion for J-spectral factorizations and to find versions of the positive real and bounded real lemmas. After that we specialize the result of Section 4.4 to the case of the standard LQ Riccati equation for statically stabilizable systems.

4.5.1   J-Spectral Factorizations

One important special case of Theorem 4.3.2 is the application to J-spectral factorizations for systems that are strongly stable. In this case, the coercivity assumption (4.4) is not satisfied.

Define for Hilbert spaces Z_+, Z_- and a real $\gamma > 0$,

$$J_{Z_+, Z_-}(\gamma) = \begin{bmatrix} I_{Z_+} & 0 \\ 0 & -\gamma^2 I_{Z_-} \end{bmatrix}.$$

Definition 4.5.1 *Suppose that $G \in \mathbf{L}_\infty(\mathcal{L}(U, Y))$, where $U = U_+ \times U_-$, $Y = Y_+ \times U_-$ and U_+, Y_+, U_- are separable Hilbert spaces. A J-spectral factorization of G is defined by*

$$G^\sim(j\omega) J_{Y_+, U_-}(\gamma) G(j\omega) = W^\sim(j\omega) J_{U_+, U_-}(\gamma) W(j\omega), \qquad (4.18)$$

where W and $W^{-1} \in \mathbf{H}_\infty(\mathcal{L}(U))$ and W and W^{-1} are regular.

Theorem 4.5.2 *Consider the strongly stable system $\Sigma(A, B, C, D)$ and let $U = U_+ \times U_-$ be its input space, $Y = Y_+ \times U_-$ be its output space and $G(s)$ be its transfer function. G has a J-spectral factorization if and only if*

1. *there exists a boundedly invertible operator $W_\infty \in \mathcal{L}(Y_+ \times U_-)$ such that*

$$D^* J_{Y_+, U_-}(\gamma) D = W_\infty^* J_{U_+, U_-}(\gamma) W_\infty \qquad (4.19)$$

and

2. *the Riccati equation (4.1) with the operators $R = D^* J_{Y_+, U_-}(\gamma) D$, $Q = J_{Y_+, U_-}(\gamma)$ and $N = D^* J_{Y_+, U_-}(\gamma)$ has a unique, self-adjoint, strongly stabilizing solution X.*

In this case, a J-spectral factor for G is given by

$$W(s) = W_\infty + L(sI - A)^{-1} B,$$

where $L = J_{U_+, U_-}(\gamma)^{-1} W_\infty^{-*}(B^* X + D^* J_{Y_+, U_-}(\gamma) C)$.

Proof For convenience of notation we write $J_U = J_{U_+, U_-}(\gamma)$ and $J_Y = J_{Y_+, U_-}(\gamma)$.

Sufficiency: The formula for $W(s)$ follows by direct substitution. It remains to prove that the spectral factor W and its inverse are in $\mathbf{H}_\infty(\mathcal{L}(U))$. It is easier to consider a normalized factor $\tilde{W}$ instead. We define $\tilde{W}(s) = W_\infty^{-1} W(s)$ and show that $\tilde{W}$ and its inverse are in $\mathbf{H}_\infty(\mathcal{L}(U))$. The Riccati feedback F_X is given by $F_X = -(W_\infty^* J_U W_\infty)^{-1}(B^* X + D^* J_Y C)$ and so we can express $\tilde{W}$ and its inverse in terms of F_X as

$$\tilde{W}(s) = I - F_X(sI - A)^{-1} B,$$
$$\tilde{W}^{-1}(s) = I + F_X(sI - A_{F_X})^{-1} B.$$

Because X is a strongly stabilizing solution, the above expression implies that $\tilde{W}^{-1} \in \mathbf{H}_\infty(\mathcal{L}(U))$. By the input-output stability of the system $\Sigma(A, B, C, D)$, we have that $G^\sim J_Y G \in \mathbf{L}_\infty(\mathcal{L}(U))$ and consequently that

$$\tilde{W}^\sim(j\omega) = G^\sim(j\omega) J_Y G(j\omega) W^{-1}(j\omega) J_U W_\infty^{-1} \in \mathbf{L}_\infty(\mathcal{L}(U)).$$

The input stability of $\Sigma(A, B, C, D)$ allows one to conclude that

$$(I - \tilde{W}^\sim(-s)) u = B^*(sI - A^*)^{-1} F_X^* u \in \mathbf{H}_2(U) \quad \text{for all } u \in U.$$

From Lemma 4.2.4 it now follows that $\tilde{W}(s)$ and hence $W(s) \in \mathbf{H}_\infty(\mathcal{L}(U))$.

Necessity: Conversely, suppose that G has a J-spectral factorization. We first prove part 1. Let the limit of the spectral factor W along the positive real axis be

$$\lim_{\lambda \to \infty} W(\lambda) = W_\infty.$$

Note that in the definition of a J-spectral factorization, we demanded that W be regular. Hence, this limit exists. Then, taking limits on both sides in (4.18), we obtain $D^* J_Y D = W_\infty^* J_U W_\infty$.

For the proof of part 2, we define $\Pi = G^\sim J_Y G$. Note that Π is the Popov function associated with the Riccati equation (4.1) with $Q = J_Y$, $N = D^* J_Y$, $R = D^* J_Y D$. Setting $M(s) = W(s)$, $N(s) = J_U W(s)$, we obtain a Wiener–Hopf factorization of Π. Consequently, the Riccati equation has a unique, self-adjoint, strongly stabilizing solution. $\square$

4.5.2 The Strict Positive Real Lemma

Let $G(s) = D + C(sI - A)^{-1}B$ be the transfer function of the strongly stable system $\Sigma(A, B, C, D)$. The strict positive real lemma states the following: G is strictly positive real (see Definition 2.2.3) if and only if $D + D^* \geq \varepsilon I$ for some $\varepsilon > 0$ and there exists a unique, self-adjoint, nonnegative, strongly stabilizing solution X to the Riccati equation

$$A^*Xz + XAz + (B^*X - C)^*(D + D^*)^{-1}(B^*X - C)z = 0 \qquad (4.20)$$

for $z \in \mathcal{D}(A)$. In this case, there exists a spectral factorization for $G + G^\sim$ (i.e., there exists $M(s) \in \mathbf{H}_\infty(\mathcal{L}(U))$ such that $M^{-1}(s) \in \mathbf{H}_\infty(\mathcal{L}(U))$ and $G + G^\sim = M^\sim M$ almost everywhere on the imaginary axis). $M(s)$ is given by

$$M(s) = (D + D^*)^{\frac{1}{2}} - (D + D^*)^{-\frac{1}{2}}(B^*X - C)(sI - A)^{-1}B.$$

Note that taking the limit along the positive real axis for $s \to \infty$ in the inequality $G(s) + G(s)^* \geq \varepsilon I$ shows that the direct feedthrough D of a strictly positive real transfer function $G(s) = D + C(sI - A)^{-1}B$ always satisfies $D + D^* \geq \varepsilon I$.

The strict positive real lemma is a direct consequence of Corollary 4.3.3, because $G(j\omega) + G(j\omega)^*$ is the Popov function corresponding to

$$A^*Xz + XAz - (B^*X + C)^*(D + D^*)^{-1}(B^*X + C)z = 0 \qquad (4.21)$$

and X is a self-adjoint stabilizing solution of (4.20) if and only if $-X$ is a self-adjoint stabilizing solution of (4.21). We still need to prove that $X \geq 0$. Note that the Riccati equation (4.20) corresponds to $N = -I$, $Q = 0$ and $R = -D - D^*$. Using (4.12), we obtain

$$X = -\Psi^*\mathcal{R}^{-1}\Psi.$$

Now, since G is strictly positive real, we have $\mathbb{F} + \mathbb{F}^* \geq \varepsilon I$, where $\mathbb{F}$ is the extended input-output map of $\Sigma(A, B, C, D)$. So, $\mathcal{R} = -\mathbb{F} - \mathbb{F}^* \leq -\varepsilon I$ and $X \geq 0$. Note that under the extra assumption that the system is approximately observable (i.e., $\ker(\Psi) = \{0\}$) we obtain that $X > 0$.

4.5.3 The Strict Bounded Real Lemma

A transfer function $G(s)$ is *bounded real* if

1. $\overline{G(\bar{s})} = G(s)$,

2. $G(s)$ is holomorphic on $\mathbb{C}_0^+$, and

3. $I - G^*(s)G(s) \geq 0$ on $\overline{\mathbb{C}_0^+}$.

When the inequality is strengthened to $I - G^*(s)G(s) \geq \varepsilon I$ for some $\varepsilon > 0$, then $G(s)$ is *strictly bounded real*. Note that if $\Sigma(A, B, C, D)$ is strictly bounded real, then automatically also $I - D^*D \geq \varepsilon I$. This follows by taking the limit as $s \to \infty$ along the positive real axis in the inequality $I - G^*(s)G(s) \geq \varepsilon I$.

Let $\Sigma(A, B, C, D)$ be a strongly stable system with transfer function $G(s)$. The strict bounded real lemma then reads: G is strictly bounded real if and only if $I - D^*D \geq \varepsilon I$ for some $\varepsilon > 0$ and there exists a unique, self-adjoint, strongly stabilizing solution X to

$$A^*Xz + XAz + (B^*X + D^*C)^*(I - D^*D)^{-1}(B^*X + D^*C)z + C^*Cz = 0 \qquad (4.22)$$

for $z \in \mathcal{D}(A)$. In this case there exists a spectral factorization $I - G(j\omega)^*G(j\omega) = M(j\omega)^*M(j\omega)$, which is given by

$$M(s) = (I - D^*D)^{\frac{1}{2}} - (I - D^*D)^{-\frac{1}{2}}(B^*X + D^*C)(sI - A)^{-1}B.$$

Furthermore, in this case $X \geq 0$ and if $\Sigma(A, -, C)$ is approximately observable, then X even satisfies $X > 0$.

The strict bounded real lemma is a direct consequence of Corollary 4.3.3, because $I - G(j\omega)^*G(j\omega)$ is the Popov function associated with

$$A^*Xz + XAz - (B^*X - D^*C)^*(I - D^*D)^{-1}(B^*X - D^*C)z - C^*Cz = 0 \quad (4.23)$$

and X is a stabilizing solution of (4.22) if and only if $-X$ is a stabilizing solution of (4.23), in a similar way as for the strict positive real lemma. That $X \geq 0$ follows from the construction in (4.12): $X = \Psi^*(I + \mathbb{F}(I - D^*D)^{-1}\mathbb{F}^*)\Psi$, where $\mathbb{F}$ is the extended input-output map of $\Sigma(A, B, C, D)$. Clearly, $X \geq 0$ by the positivity of $I - D^*D$. Note that $X > 0$ if $\Sigma(A, -, C)$ is approximately observable.

4.5.4 The Coercive LQ Case

Let $\Sigma(A, B, C)$ be a statically stabilizable system and let

$$\begin{bmatrix} Q & N^* \\ N & R \end{bmatrix} \geq \alpha I \qquad (4.24)$$

for some $\alpha > 0$. This coercivity assumption is satisfied, for instance, in the standard LQ case, where $N = 0$, $Q = I$, $R = I$, or in the case that $N = 0$, $R \geq \delta I$ and $Q \geq \mu I$ for some $\mu > 0$, $\delta > 0$. Under these assumptions, clearly we have $\mathcal{F}(y, u) \geq \alpha(\|y\|^2 + \|u\|^2)$. Hence, by part 2 of Theorem 4.4.3, the corresponding Riccati equation will have a unique, self-adjoint, strongly stabilizing solution X.

As an application of this result, we look at the standard LQ control problem for a statically stabilizable system with direct feedthrough.

Theorem 4.5.3 *Consider the standard LQ control problem for the statically stabilizable system $\Sigma(A, B, C, D)$. That is, consider the problem of minimizing*

$$\mathcal{J}(u, z_0) = \int_0^\infty \|y(t)\|^2 + \|u(t)\|^2 dt \qquad (4.25)$$

subject to

$$\dot{z}(t) = Az(t) + Bu(t), \quad z(0) = z_0,$$
$$y(t) = Cz(t) + Du(t).$$

The Riccati equation associated with this problem is given by

$$A^*Xz + XAz - (XB + C^*D)(I + D^*D)^{-1}(B^*X + D^*C)z + C^*Cz = 0 \quad (4.26)$$

for $z \in \mathcal{D}(A)$, and it has a unique, self-adjoint, strongly stabilizing solution.

Proof For this LQ problem, $Q = I$, $N = D^*$, $R = I + D^*D$ and it follows immediately that (4.26) is the corresponding Riccati equation.

Note that, even though Q and R are coercive, in general

$$\begin{bmatrix} Q & N^* \\ N & R \end{bmatrix} = \begin{bmatrix} I & D \\ D^* & I + D^*D \end{bmatrix}$$

is not coercive. However, we can transform the problem into an equivalent one without cross-terms in the cost and for a system without direct feedthrough. To do so, we define $\tilde{y} = (I + DD^*)^{-\frac{1}{2}}Cz$ and apply the state feedback

$$u(t) = -(I + D^*D)^{-1}D^*Cz + (I + D^*D)^{-\frac{1}{2}}\tilde{u}.$$

Thus, we obtain the following equivalent LQ control problem:

$$\tilde{J}(\tilde{u}, z_0) = \int_0^\infty \|\tilde{y}\|^2 + \|\tilde{u}\|^2 dt$$

subject to

$$\dot{z}(t) = \tilde{A}z(t) + \tilde{B}\tilde{u}(t), \quad z(0) = z_0,$$
$$\tilde{y}(t) = \tilde{C}z(t),$$

where $\tilde{A} = A - B(I + D^*D)^{-1}D^*C$, $\tilde{B} = B(I + D^*D)^{-\frac{1}{2}}$ and $\tilde{C} = (I + DD^*)^{-\frac{1}{2}}C$. It can easily be checked that $\Sigma(\tilde{A}, \tilde{B}, \tilde{C})$ inherits the static stabilizability property from $\Sigma(A, B, C)$. In this new LQ problem, $Q = I$, $R = I$ and $N = 0$, so condition (4.24) is satisfied and the Riccati equation

$$\tilde{A}^*Xz + X\tilde{A}z - X\tilde{B}\tilde{B}^*Xz + \tilde{C}^*\tilde{C}z = 0 \quad \text{for } z \in \mathcal{D}(A) \tag{4.27}$$

has a unique, self-adjoint, nonnegative, strongly stabilizing solution. From the fact that $(I + DD^*)^{-1} = I + D(I + D^*D)^{-1}D^*$ it is easy to see that the two Riccati equations (4.26) and (4.27) are exactly the same. Hence, (4.26) also has a unique, self-adjoint, nonnegative, strongly stabilizing solution. □

Dually, it can be shown that if $\Sigma(A^*, C^*, B^*, D^*)$ is statically stabilizable, then the standard filter Riccati equation for $z \in \mathcal{D}(A^*)$,

$$AYz + YA^*z - (YC + BD^*)(I + DD^*)^{-1}(CY + DB)z + BB^*z = 0, \tag{4.28}$$

has a unique, self-adjoint, nonnegative, strongly stabilizing solution. Similar to what was presented in Section 3.3, this means that the associated closed-loop system $\Sigma(A^* + C^*L_Y^*, C^*, [\begin{array}{cc} B & L_Y \end{array}]^*, D^*)$ is a strongly stable system. For filtering, one would be interested in the stability of the system $\Sigma(A + L_YC, [\begin{array}{cc} B & L_Y \end{array}], C, D)$ and we do obtain input stability, output stability and input-output stability for this system. However, in general $A + L_YC$ will not generate a strongly stable semigroup, although its adjoint does.

We should remark that the result in this section is weaker than the result in Theorem 3.3.1. In the same way as here, we could include a direct feedthrough term into that result. Hence, the only difference between the two results is that we need to impose static stabilizability here, which is stronger than the combination of strong stabilizability and strong detectability that was assumed in Theorem 3.3.2.

4.6 Examples

In this section, we will illustrate the results of the preceding sections by applying them to a number of problems. Rather than considering the most general class of systems possible, we specialize to systems of the $\Sigma(A, B, B^*)$ class, which was introduced in Section 2.2. It will turn out that the conditions for the existence of solutions of the Riccati equations can in many cases be easily checked for this class of systems. We will consider Riccati equations for systems $\Sigma(A, B, B^*, D)$ under the assumptions A1–A5 (see page 17).

4.6.1 The Strict Positive Real Lemma

We consider the strict positive real lemma for systems $\Sigma(A_B, B, B^*, D)$, where $A_B = A - BB^*$ and $D + D^* \geq \varepsilon I$ for some $\varepsilon > 0$ under the assumptions A1–A5. In this case, the transfer function $G(s) = D + B^*(sI - A_B)^{-1}B$ is strictly positive real. Therefore, by the result of Section 4.5.2, the Riccati equation for $z \in \mathcal{D}(A)$,

$$A_B^* X z + X A_B z + (XB - B)(D + D^*)^{-1}(B^* X - B^*)z = 0,$$

has a unique, self-adjoint, strongly stabilizing solution X.

Furthermore, there exists a spectral factorization

$$G(j\omega) + G(j\omega)^* = M(j\omega)^* M(j\omega), \tag{4.29}$$

where the spectral factor M is given by

$$M(s) = (D + D^*)^{\frac{1}{2}} - (D + D^*)^{-\frac{1}{2}}(B^* X - B^*)(sI - A_B)^{-1}B.$$

4.6.2 The Standard LQ Riccati Equation

For the system $\Sigma(A, B, B^*, D)$ satisfying assumptions A1–A5, we consider the standard LQ optimal control problem, i.e., we want to minimize the criterion (4.25). From the example in Section 4.5.4, it follows that the associated Riccati equation for $z \in \mathcal{D}(A)$,

$$A^* X z + X A z - (XB + BD)(I + D^* D)^{-1}(B^* X + D^* B^*)z + BB^* z = 0, \tag{4.30}$$

has a unique, strongly stabilizing solution X. This case was also studied in Curtain and Zwart [37] for the same class of systems.

Chapter 5

Numerical Approximation

Controllers designed on the basis of an infinite-dimensional model of the plant are very often infinite-dimensional, as well. In particular, this is the case for LQ-, LQG- and H_∞-controllers. Of course, when one wants to actually compute or even implement such a controller, one has to approximate it by a finite-dimensional controller. Therefore, over the years much research has been directed toward developing approximation schemes for these controllers and, in particular, approximation schemes for numerical solutions of algebraic Riccati equations. Many of the papers use an approach based on approximation results for C_0-semigroups, i.e., they use a version of the Trotter–Kato theorem in the proof of convergence of the solution of the Riccati equation (see, for instance, Banks and Burns [11], Gibson [49], Ito [55], Kappel and Salamon [58] and the references therein). In all of these papers it was assumed that A generates an exponentially stable semigroup or that the pair (A, B) is exponentially stabilizable. Therefore, after the results on the existence of the strongly stabilizing solution of Riccati equations in Chapters 3 and 4, it is logical to next consider the approximation of those strongly stabilizing solutions by finite-dimensional ones. In this chapter we restrict ourselves to the approximation of the strongly stabilizing solution of the standard LQ Riccati equation.

5.1 The LQ Riccati Equation for Strongly Stabilizable Systems

As in Chapter 3, we consider the following optimal control problem: For the system

$$\dot{z}(t) = Az(t) + Bu(t), \quad z(0) = z_0,$$
$$y(t) = Cz(t),$$
(5.1)

we want to find an input $\bar{u} \in \mathbf{L}_2(0, \infty; U)$ that minimizes the quadratic cost criterion

$$\mathcal{J}(u, z_0) = \int_0^\infty (\|y(t)\|^2 + \|u(t)\|^2)dt.$$
(5.2)

In Theorem 3.3.2, we proved that strong stabilizability of $\Sigma(A, B, C)$ is sufficient for the existence of a minimizing control $\bar{u}$ and a self-adjoint, nonnegative solution to the algebraic Riccati equation

$$A^*Xz + XAz - XBB^*Xz + CC^*z = 0$$
(5.3)

for $z \in \mathcal{D}(A)$. In the same theorem, it was proven that if, in addition, $\Sigma(A, B, C)$ is strongly detectable, then the solution X is a strongly stabilizing solution of (5.3) and is the only self-adjoint solution with this property.

Next we show how to reformulate the minimization problem for a strongly stabilizable system as an equivalent one for a strongly stable system. We do this in a different way from that in Chapter 4 and, as a result, we obtain a different formula for the strongly stabilizing solution X. The conditions that we impose this time on stabilizability and detectability of the system are stronger than those in Chapter 3, but weaker than those in Chapter 4. We elaborate on the differences at the end of this section.

Theorem 5.1.1 *Consider the problem of minimizing the cost functional (5.2) subject to $\Sigma(A, B, C)$ under the assumption that there exist $F \in \mathcal{L}(Z, U)$ and $L \in \mathcal{L}(Y, Z)$ such that*

C1. *$A + BF$ and $A + LC$ generate strongly stable C_0-semigroups $T_F(t)$ and $T_L(t)$, respectively;*

C2. *$\Sigma\left(A + BF, B, \begin{bmatrix} F \\ C \end{bmatrix}\right)$ is input-output stable and output stable;*

C3. *$\Sigma(A + LC, \begin{bmatrix} L & B \end{bmatrix}, F)$ is input stable, output stable and input-output stable.*

Then this problem is equivalent to that of minimizing the cost functional

$$\mathcal{J}_F(v, z_0) = \int_0^\infty (\|v(t) + Fz(t)\|^2 + \|y(t)\|^2)dt \tag{5.4}$$

subject to

$$\dot{z}(t) = (A + BF)z(t) + Bv(t), \quad z(0) = z_0, \\ y(t) = Cz(t). \tag{5.5}$$

The optimal input for the optimal control problem (5.4), (5.5) is given by

$$\bar{v}(t) = -(\mathbb{F}_F^* \mathbb{F}_F)^{-1} \mathbb{F}_F^* \Psi_F z_0, \tag{5.6}$$

where Ψ_F and $\mathbb{F}_F$ are the extended output map and the extended input-output map, respectively, of the system

$$\Sigma\left(A + BF, B, \begin{bmatrix} F \\ C \end{bmatrix}, \begin{bmatrix} I \\ 0 \end{bmatrix}\right).$$

Moreover, there exists a unique, self-adjoint, nonnegative, strongly stabilizing solution to the algebraic Riccati equation (5.3), which is given by

$$X = \Psi_F^* \left[I - \mathbb{F}_F(\mathbb{F}_F^* \mathbb{F}_F)^{-1} \mathbb{F}_F^* \right] \Psi_F. \tag{5.7}$$

Proof (*a*) Note that with $u = v + Fz$, the systems (5.1) and (5.5) have the same solution

$$z(t) = T_F(t)z_0 + \int_0^t T_F(t-s)Bv(s)ds = T(t)z_0 + \int_0^t T(t-s)Bu(s)ds$$

and $\mathcal{J}_F(v, z_0) = \mathcal{J}(u, z_0)$. Therefore, the two optimization problems are equivalent if the following two statements are equivalent:

- $u \in \mathbf{L}_2(0, \infty; U)$ and $\mathcal{J}(u, z_0) < \infty$;

- $v = u - Fz \in \mathbf{L}_2(0, \infty; U)$ and $\mathcal{J}_F(v, z_0) < \infty$.

It follows from assumption C2 that $v \in \mathbf{L}_2(0, \infty; U)$ if and only if $\mathcal{J}_F(v, z_0) < \infty$. Hence, it suffices to prove that $v \in \mathbf{L}_2$ if and only if $u \in \mathbf{L}_2$ and $\mathcal{J}(u, z_0) < \infty$.

Suppose first that $v \in \mathbf{L}_2(0, \infty; U)$. Since $\Sigma(A + BF, B, F)$ is input-output stable, $Fz(t) - FT_F(t)z_0 \in \mathbf{L}_2(0, \infty; U)$ and since $\Sigma(A + BF, B, F)$ is output stable, $FT_F(t)z_0$ and hence $Fz(t) \in \mathbf{L}_2(0, \infty; U)$. Thus, $u \in \mathbf{L}_2(0, \infty; U)$. The output stability and input-output stability of $\Sigma(A + BF, B, C)$ imply that $Cz(t) \in \mathbf{L}_2(0, \infty; Y)$ which, together with $u \in \mathbf{L}_2(0, \infty; U)$, show that $\mathcal{J}(u, z_0) < \infty$.

Conversely, suppose that $u \in \mathbf{L}_2(0, \infty; U)$ and $\mathcal{J}(u, z_0) < \infty$ and write

$$z(t) = T_L(t)z_0 + \int_0^t T_L(t - s) \begin{bmatrix} B & -L \end{bmatrix} \begin{pmatrix} u(s) \\ y(s) \end{pmatrix} ds.$$

Then $Fz(t) - FT_L(t)z_0 \in \mathbf{L}_2(0, \infty; U)$, since $\Sigma(A + LC, \begin{bmatrix} B & L \end{bmatrix}, F)$ is input-output stable. In addition, $FT_L(t)z_0 \in \mathbf{L}_2(0, \infty; U)$, since $\Sigma(A + LC, B, F)$ is output stable. So $Fz(t) \in \mathbf{L}_2(0, \infty; U)$ and therefore also $v = u - Fz \in \mathbf{L}_2(0, \infty; U)$.

(*b*) We now obtain an explicit solution of the minimization of (5.4) for the stable system (5.5) in terms of the bounded maps $\mathbb{F}_F$, Ψ_F given by

$$\Psi_F z_0 = \begin{bmatrix} F \\ C \end{bmatrix} T_F(t)z_0,$$

$$(\mathbb{F}_F v)(t) = \begin{bmatrix} v(t) \\ 0 \end{bmatrix} + \begin{bmatrix} F \\ C \end{bmatrix} \int_0^t T_F(t - s)Bv(s)ds.$$

Since $\mathcal{J}_F(v, z_0) < \infty$ for all $v \in \mathbf{L}_2(0, \infty; U)$, we can write (5.4) equivalently as

$$\mathcal{J}_F(v, z_0) = \|\Psi_F z_0 + \mathbb{F}_F v\|^2.$$

Next, we show that the cost functional $\mathcal{J}_F$ is coercive. Let the operator $\mathbb{F}_{LF}$ denote the input-output map of the system $\Sigma(A + LC, \begin{bmatrix} B & -L \end{bmatrix}, -F, \begin{bmatrix} I & 0 \end{bmatrix})$. By assumption C3, $\mathbb{F}_{LF} \in \mathcal{L}(\mathbf{L}_2(0, \infty; U \times Y), \mathbf{L}_2(0, \infty; U))$. We show that $\mathbb{F}_{LF}\mathbb{F}_F = I_{\mathbf{L}_2(0, \infty; U)}$, using a frequency domain argument. Let $\hat{}$ denote Laplace transforms and let $v \in \mathbf{L}_2(0, \infty; U)$, $y \in \mathbf{L}_2(0, \infty; Y)$. Then,

$$(\widehat{\mathbb{F}_{LF}(v, y)})(s) = \left(\begin{bmatrix} I & 0 \end{bmatrix} - F(sI - A - LC)^{-1} \begin{bmatrix} B & -L \end{bmatrix} \right) \begin{pmatrix} \hat{v}(s) \\ \hat{y}(s) \end{pmatrix}$$

and

$$(\widehat{\mathbb{F}_F v})(s) = \left(\begin{bmatrix} I \\ 0 \end{bmatrix} + \begin{bmatrix} F \\ C \end{bmatrix} (sI - A - BF)^{-1} B \right) \hat{v}(s).$$

Now,

$$(\widehat{\mathbb{F}_{LF}\mathbb{F}_F v})(s)$$
$$= \left\{ F(sI - A - LC)^{-1}(LC - BF)(sI - A - BF)^{-1}B \right.$$
$$\left. - F(sI - A - LC)^{-1}B + F(sI - A - BF)^{-1}B + I \right\} \hat{v}(s)$$
$$= F(sI - A - LC)^{-1} \left\{ -sI + A + BF + sI - A - LC + LC - BF \right\}$$
$$\quad \cdot (sI - A - BF)^{-1}B\hat{v}(s) + \hat{v}(s)$$
$$= \hat{v}(s).$$

Consequently, $(\mathbb{F}_{LF}\mathbb{F}_F v)(t) = v(t)$ for all $v \in \mathbf{L}_2(0, \infty; U)$.

We deduce further that

$$
\begin{aligned}
\|v\|^2 &= \|\mathbb{F}_{LF}\mathbb{F}_F v\|^2 \\
&\leq \|\mathbb{F}_{LF}\|^2 \|\mathbb{F}_F v\|^2 \\
&= \|\mathbb{F}_{LF}\|^2 \langle \mathbb{F}_F v, \mathbb{F}_F v \rangle \\
&\leq \|\mathbb{F}_{LF}\|^2 \|\mathbb{F}_F^* \mathbb{F}_F v\| \|v\|.
\end{aligned}
$$

Note that, because of the identity $\mathbb{F}_{LF}\mathbb{F}_F = I$, we cannot have $\|\mathbb{F}_{LF}\| = 0$. Thus,

$$
\|\mathbb{F}_F^* \mathbb{F}_F v\| \geq \|\mathbb{F}_{LF}\|^{-2} \|v\|. \tag{5.8}
$$

We can now apply the theory of paragraphs 1.1.1–1.1.3 in Lions [69] to show that there exists a unique minimizing control $\bar{v} \in \mathbf{L}_2(0,\infty;U)$, which is characterized by

$$
\mathbb{F}_F^* \mathbb{F}_F \bar{v} + \mathbb{F}_F^* \Psi_F z_0 = 0. \tag{5.9}
$$

Because (5.8) implies that $\mathbb{F}_F^* \mathbb{F}_F$ is boundedly invertible, we can solve (5.9) for $\bar{v}$ to obtain (5.6).

From Theorem 3.3.2 we know that

$$
\mathcal{J}_F(\bar{v}, z_0) = \mathcal{J}(\bar{u}, z_0) = \langle X_0 z_0, z_0 \rangle
$$

and (5.7) follows. That X is the unique, self-adjoint, strongly stabilizing solution also follows from Theorem 3.3.2.   □

It is interesting to note that although the strategy in Section 4.4 is also to express the general control problem via a feedback in terms of an equivalent stable one, the feedbacks are different and the formulas for X are very different. In the stable case ($F = 0$, $L = 0$), however, the formulas for X coincide. For proving convergence results for the standard LQ Riccati equation the approach in this chapter is more convenient.

It turns out that the conditions C1, C2, C3 are key to proving our main approximation result in Section 5.3. Consequently, we introduce the following definition.

Definition 5.1.2 *The system $\Sigma(A, B, C)$ is* strongly stabilizable-detectable *if there exist $F \in \mathcal{L}(Z, U)$ and $L \in \mathcal{L}(Y, Z)$ such that*

C1. *$A + BF$ and $A + LC$ generate strongly stable C_0-semigroups $T_F(t)$ and $T_L(t)$, respectively;*

C2. *$\Sigma\left(A + BF, B, \begin{bmatrix} F \\ C \end{bmatrix}\right)$ is input-output stable and output stable;*

C3. *$\Sigma(A + LC, [\,L\ \ B\,], F)$ is input-output stable, input stable and output stable.*

Let us point out the difference between a system that is strongly stabilizable and strongly detectable and a system that is strongly stabilizable-detectable. As a short inspection reveals, every strongly stabilizable-detectable system is both strongly stabilizable and strongly detectable; the only candidates for F and L in Definition 5.1.2 are F's which strongly stabilize the system and L's which strongly detect the system, in the sense of Definition 2.1.7. However, the operators L and F have to be related to each other in some sense: in condition C3 of Definition 5.1.2, both F and L play a role. The output stability and input-output stability requirements will put a restriction on the possible combinations of F and L. Second, the input-output stability in condition C2 is, in general, not satisfied if F only strongly stabilizes the system. Note, however, that the concept of strong stabilizability-detectability is a weaker notion than that of static stabilizability.

We remark that if both $A + BF$ and $A + LC$ generate exponentially stable C_0-semi-groups on Z, then the conditions C1, C2 and C3, above, hold automatically.

5.2 Approximation of Strongly Stable Systems

In this section we show how to approximate a strongly stable system by a sequence of finite-dimensional systems. We show that our approximation scheme guarantees the convergence of the extended output maps and the extended input-output maps. In Section 5.3, we use these results to prove the strong convergence of a sequence of stabilizing solutions of finite-dimensional Riccati equations to the strongly stabilizing solution of the infinite-dimensional standard LQ Riccati equation.

Consider again the system on the Hilbert spaces Z, U, Y,

$$
\begin{aligned}
\dot{z}(t) &= Az(t) + Bu(t), \\
y(t) &= Cz(t).
\end{aligned}
\tag{5.10}
$$

This infinite-dimensional system is approximated by a sequence of systems on the finite-dimensional spaces $Z^N = \mathbb{R}^{k(N)}, U^N = \mathbb{R}^{m(N)}, Y^N = \mathbb{R}^{p(N)}$,

$$
\begin{aligned}
\dot{z}^N(t) &= A^N z^N(t) + B^N u^N(t), \\
y^N(t) &= C^N z^N(t).
\end{aligned}
\tag{5.11}
$$

We assume the existence of injective linear maps

$$
\begin{aligned}
i^N &: \mathbb{R}^{k(N)} \to Z, \\
j^N &: \mathbb{R}^{m(N)} \to U, \\
k^N &: \mathbb{R}^{p(N)} \to Y
\end{aligned}
$$

and surjective linear maps

$$
\begin{aligned}
\pi^N &: Z \to \mathbb{R}^{k(N)}, \\
\rho^N &: U \to \mathbb{R}^{m(N)}, \\
\sigma^N &: Y \to \mathbb{R}^{p(N)}
\end{aligned}
$$

such that $\pi^N i^N$, $\rho^N j^N$ and $\sigma^N k^N$ are identity maps and $i^N \pi^N$, $j^N \rho^N$ and $k^N \sigma^N$ are orthogonal projections. Note that it is not necessary that $(i^N)^* = \pi^N$, $(j^N)^* = \rho^N$ or $(k^N)^* = \sigma^N$. On the spaces $\mathbb{R}^{k(N)}, \mathbb{R}^{m(N)}$ and $\mathbb{R}^{p(N)}$ we will always consider the induced inner products

$$
\begin{aligned}
\langle z_1, z_2 \rangle_{k(N)} &= \langle i^N z_1, i^N z_2 \rangle_Z, \\
\langle u_1, u_2 \rangle_{m(N)} &= \langle j^N u_1, j^N u_2 \rangle_U, \\
\langle y_1, y_2 \rangle_{p(N)} &= \langle j^N y_1, k^N y_2 \rangle_Y.
\end{aligned}
$$

$(A^N)^*, (B^N)^*, (C^N)^*$ will denote the adjoint matrices with respect to the induced inner products. Because we always use the induced inner products on $\mathbb{R}^{k(N)}, \mathbb{R}^{m(N)}$ and $\mathbb{R}^{p(N)}$, it is obvious that $\|i^N\| = \|j^N\| = \|k^N\| = \|\pi^N\| = \|\rho^N\| = \|\sigma^N\| = 1$. Define $Z^N = \text{range}(i^N\pi^N)$, $U^N = \text{range}(j^N\rho^N)$ and $Y^N = \text{range}(k^N\sigma^N)$. Then,

$$\|i^N B^N \rho^N\|_{\mathcal{L}(U^N,H^N)} = \|B^N\|_{\mathcal{L}(\mathbb{R}^{m(N)},\mathbb{R}^{k(N)})},$$
$$\|k^N C^N \pi^N\|_{\mathcal{L}(H^N,Y^N)} = \|C^N\|_{\mathcal{L}(\mathbb{R}^{k(N)},\mathbb{R}^{p(N)})},$$

and similar equalities hold for all operators defined in an analogous way.

Let $T(\cdot), \tilde{\Phi}, \Psi, \mathbb{F}, G(s)$ be the semigroup, extended input map, extended output map, input-output map and transfer function, respectively, associated with $\Sigma(A,B,C)$, and let $T^N(\cdot), \tilde{\Phi}^N, \Psi^N, \mathbb{F}^N, G^N(s)$ be the same operators but associated with the system $\Sigma(A^N, B^N, C^N)$. In the latter case, we are dealing with finite-dimensional systems. Hence, $T^N(t) = e^{A^N t}$ and $(\Psi^N z)(t) = C^N e^{A^N t} z$. Let us define the notion of strong convergence of a sequence of systems.

Definition 5.2.1 *A sequence of systems $\Sigma(A^N, B^N, C^N)$ converges strongly to the system $\Sigma(A,B,C)$ if, for all $z \in Z$,*

$$T(t)z = \lim_{N\to\infty} i^N e^{A^N t} \pi^N z,$$
$$T^*(t)z = \lim_{N\to\infty} i^N e^{(A^N)^* t} \pi^N z$$

uniformly on compact time intervals and, for all $z \in Z$, $u \in U$, $y \in Y$,

$$Bu = \lim_{N\to\infty} i^N B^N \rho^N u,$$
$$B^* z = \lim_{N\to\infty} j^N (B^N)^* \pi^N z,$$
$$Cz = \lim_{N\to\infty} k^N C^N \pi^N z,$$
$$C^* y = \lim_{N\to\infty} i^N (C^N)^* \sigma^N y,$$
$$u = \lim_{N\to\infty} j^N \rho^N u,$$
$$y = \lim_{N\to\infty} k^N \sigma^N y.$$

Note that the convergence of the semigroup in the above definition implies, by taking $t = 0$, that for all $z \in Z$,

$$z = \lim_{N\to\infty} i^N \pi^N z.$$

In connection with the convergence of a sequence of finite-dimensional systems to an infinite-dimensional system, we introduce a notion of uniform stability of the sequence of finite-dimensional systems.

Definition 5.2.2 *A sequence of systems $\Sigma(A^N, B^N, C^N)$ is uniformly output stable if there exists an $M > 0$ such that for all $z \in Z$,*

$$\sup_N \int_0^\infty \|k^N C^N e^{A^N t} \pi^N z\|^2 = \sup_N \int_0^\infty \|k^N (\Psi^N \pi^N z)(t)\|^2 dt$$
$$\leq M\|z\|^2.$$

A sequence of systems $\Sigma(A^N, B^N, C^N)$ *is* uniformly input-output stable *if there exists an* $M > 0$ *such that*

$$\sup_N \|k^N \mathbb{F}^N \rho^N\| = \sup_N \|k^N G^N(s) \rho^N\|_{H_\infty}$$

$$= \sup_N \|k^N C^N (sI - A^N)^{-1} B^N \rho^N\|_{H_\infty} \le M.$$

In Lemmas 5.2.4 and 5.2.6, we show the conditions under which the extended output maps and extended input-output maps of a sequence of finite-dimensional systems converge to their infinite-dimensional counterparts. In the proof of these lemmas, as well as in all other convergence proofs in this chapter, we will use the fact that if the sequences of operators R^N and S^N converge strongly to R and S, respectively, then $R^N S^N$ converges strongly to RS. Often, in our convergence results we will have to prove the convergence of a sequence of operators, operating on $L_2(0, \infty)$. In these cases, the following lemma allows us to restrict our attention to the dense subset of L_2 functions with compact support.

Lemma 5.2.3 *Let* X_1 *and* X_2 *be Banach spaces and let* $R_n \in \mathcal{L}(X_1, X_2)$ *for all* $n \in \mathbb{N}$. *Let* S *be a dense subset of* X_1 *and let* $R \in \mathcal{L}(X_1, X_2)$. *Assume that*

1. *there exists an* $M > 0$ *such that for all* n, $\|R_n\| \le M$, *and*

2. $\lim\limits_{n \to \infty} R_n x = Rx$ *for all* $x \in S$.

Then, $\lim\limits_{n \to \infty} R_n x = Rx$ *for all* $x \in X_1$.

The proof of this lemma is straightforward, and hence it is omitted.

Lemma 5.2.4 *Suppose that* $T(t)$ *is strongly stable,* $\Sigma(A, B, C)$ *is output stable, the sequence of finite-dimensional systems* $\Sigma(A^N, B^N, C^N)$ *converges strongly to* $\Sigma(A, B, C)$ *and* $\Sigma(A^N, B^N, C^N)$ *is uniformly output stable. Then, for all* $z \in Z$,

$$\lim_{N \to \infty} k^N \Psi^N \pi^N z = \Psi z$$

in $\mathbf{L}_2(0, \infty; Y)$ *and*

$$\lim_{N \to \infty} i^N (\Psi^N)^* \sigma^N y = \Psi^* y$$

in Z *for all* $y \in \mathbf{L}_2(0, \infty; Y)$.

Proof For any $t_1 > 0$, we obtain

$$\|\Psi z - k^N \Psi^N \pi^N z\|^2_{L_2(0,\infty;Y)}$$

$$\le \int_0^{t_1} \|CT(t)z - k^N C^N e^{A^N t} \pi^N z\|^2 dt + 3 \int_0^\infty \|CT(t_1 + t)z\|^2 dt$$

$$+ 3 \int_0^\infty \|k^N C^N e^{A^N t} (e^{A^N t_1} \pi^N z - \pi^N T(t_1)z)\|^2 dt$$

$$+ 3 \int_0^\infty \|k^N C^N e^{A^N t} \pi^N T(t_1)z\|^2 dt$$

$$=: \alpha_1 + \alpha_2 + \alpha_3 + \alpha_4.$$

Now we derive estimates for each of the integral terms $\alpha_1, \alpha_2, \alpha_3, \alpha_4$. For α_1, we have

$$\alpha_1 \le 2 \int_0^{t_1} \|(C - k^N C^N \pi^N) T(t)z\|^2 dt$$

$$+ 2 \sup_N \|k^N C^N \pi^N\|^2 \int_0^{t_1} \|T(t)z - i^N e^{A^N t} \pi^N z\|^2 dt.$$

The right-hand side tends to zero as N tends to infinity by the Lebesgue dominated convergence theorem because the systems $\Sigma(A^N, B^N, C^N)$ are strongly convergent to $\Sigma(A, B, C)$. α_2 satisfies

$$\alpha_2 = 3 \int_0^\infty \|CT(t_1 + t)z\|^2 dt = 3\|\Psi T(t_1)z\|^2 \leq 3\|\Psi\|^2 \|T(t_1)z\|^2,$$

which tends to zero as t_1 tends to infinity, independently of N, because of the output stability of $\Sigma(A, B, C)$ and the strong stability of $T(t)$. Since the systems $\Sigma(A^N, B^N, C^N)$ are uniformly output stable, there exists a positive constant M such that

$$\alpha_3 \leq 3M \|i^N e^{A^N t_1} \pi^N z - T(t_1)z\|^2.$$

The right-hand side tends to zero as N tends to infinity because $\Sigma(A^N, B^N, C^N)$ converges strongly to $\Sigma(A, B, C)$. For the same constant M, we have

$$\alpha_4 \leq 3M \|T(t_1)z\|^2,$$

the right-hand side of which tends to zero, independently of N, because $T(t)$ is strongly stable. These estimates together show that $k^N \Psi^N \pi^N z \to \Psi z$ strongly as $N \to \infty$. Because $\|i^N (\Psi^N)^* \sigma^N\| = \|\Psi^N\| \leq M$, it follows from Lemma 5.2.3 that it is sufficient for the proof of convergence of $i^N (\Psi^N)^* \sigma^N y$ to $\Psi^* y$ to consider $y \in \mathbf{L}_2(0, \infty; Y)$ with compact support. So, let $y \in \mathbf{L}_2(0, \infty; Y)$ have support on $[0, t_1]$, $t_1 > 0$. Then,

$$\|\Psi^* y - i^N (\Psi^N)^* \sigma^N y\| \leq \int_0^{t_1} \|(T^*(t) - i^N e^{(A^N)^* t} \pi^N) C^* y(t)\| dt$$

$$+ \int_0^{t_1} \|i^N e^{(A^N)^* t} \pi^N\| \|(C^* - i^N (C^N)^* \sigma^N) y(t)\| dt.$$

Both integrals tend to zero as $N \to \infty$ because the systems $\Sigma(A^N, B^N, C^N)$ are strongly convergent to $\Sigma(A, B, C)$. $\qquad\square$

Corollary 5.2.5 *Under the assumptions of Lemma 5.2.4,*

$$\lim_{N \to \infty} \|k^N C^N (sI - A^N)^{-1} \pi^N z - C(sI - A)^{-1} z\|_{H_2(Y)} = 0$$

for all $z \in Z$.

Proof This result follows directly by taking Laplace transforms in Lemma 5.2.4. $\qquad\square$

Lemma 5.2.6 *Suppose that the assumptions of Lemma 5.2.4 hold and, in addition, $\Sigma(A, B, C)$ is input-output stable and $\Sigma(A^N, B^N, C^N)$ is uniformly input-output stable. Then, for all $u \in \mathbf{L}_2(0, \infty; U)$,*

$$\lim_{N \to \infty} k^N \mathbb{F}^N \rho^N u = \mathbb{F} u$$

in $\mathbf{L}_2(0, \infty; Y)$. Similarly, for all $y \in \mathbf{L}_2(0, \infty; Y)$,

$$\lim_{N \to \infty} j^N (\mathbb{F}^N)^* \sigma^N y = \mathbb{F}^* y$$

in $\mathbf{L}_2(0, \infty; U)$.

Proof By the uniform input-output stability of the systems $\Sigma(A^N, B^N, C^N)$, the operators $k^N \mathbb{F}^N \rho^N$ are uniformly bounded in N. Therefore, by Lemma 5.2.3 it is sufficient to prove the convergence $\mathbb{F}^N u \to \mathbb{F}u$ for input functions u with compact support. Let $u \in \mathbf{L}_2(0,\infty;U)$ with support in $[0,t_1]$, let y, y^N be defined by $y = \mathbb{F}u$, $y^N = k^N \mathbb{F}^N \rho^N u$ and define $M_1 = \sup_N \sup_{t \in [0,t_1]} \|i^N e^{A^N t} \pi^N\|$. Then,

$$\|y(t) - y^N(t)\|^2 \leq 3 \int_0^t \|(C - k^N C^N \pi^N) T(s) Bu(t-s)\|^2 ds$$

$$+ 3 \sup_N \|k^N C^N \pi^N\| \int_0^t \|(T(s) - i^N e^{A^N s} \pi^N) Bu(t-s)\|^2 ds$$

$$+ 3 M_1 \sup_N \|k^N C^N \pi^N\| \int_0^t \|(B - i^N B^N \rho^N) u(t-s)\|^2 ds$$

for $0 \leq t \leq t_1$. By the strong convergence of $\Sigma(A^N, B^N, C^N)$ to $\Sigma(A, B, C)$ and the Lebesgue dominated convergence theorem, it now follows in the same way as in Kappel and Salamon [58] that $\|y - y^N\|_{L_2(0,t_1;Y)} \to 0$ as $N \to \infty$. Furthermore, we define

$$z_u = \int_0^{t_1} T(t_1 - s) Bu(s) ds,$$

$$z_u^N = \int_0^{t_1} e^{A^N(t_1 - s)} B^N \rho^N u(s) ds.$$

By the strong convergence of $\Sigma(A^N, B^N, C^N)$ to $\Sigma(A, B, C)$, we have that $z_u^N \to z_u$ as $N \to \infty$. In addition, since u has support in $[0,t_1]$, we obtain for $t \geq t_1$,

$$y(t) = (\Psi z_u)(t - t_1),$$
$$y^N(t) = k^N (\Psi^N z_u^N)(t - t_1).$$

Lemma 5.2.4 is now used to prove that $\|y - y^N\|_{L_2(t_1,\infty;Y)} \to 0$ as $N \to \infty$.

To prove the convergence of $(\mathbb{F}^N)^*$, note that because of the uniform boundedness of $(\mathbb{F}^N)^*$ it is again sufficient to only consider y with compact support $[0,t_1]$, by Lemma 5.2.3. The proof then proceeds completely analogously to the case above. □

Corollary 5.2.7 *Under the assumptions of Lemma 5.2.6,*

$$\lim_{N \to \infty} \|k^N C^N (sI - A^N)^{-1} B^N \rho^N \hat{u}(s) - C(sI - A)^{-1} B\hat{u}(s)\|_{H_2(Y)} = 0$$

for all $\hat{u}(s) \in \mathbf{H}_2(U)$. Similarly, for all $\hat{y}(s) \in \mathbf{H}_2(Y)$,

$$\lim_{N \to \infty} \|j^N (B^N)^* (sI - (A^N)^*)^{-1} (C^N)^* \sigma^N \hat{y}(s) - B^*(sI - A^*)^{-1} C^* \hat{y}(s)\|_{\mathbf{H}_2(U)} = 0.$$

Proof This result follows directly by taking Laplace transforms in Lemma 5.2.6. □

We remark that the above properties even hold for $\hat{u}(s) \equiv u$, $\hat{y}(s) \equiv y$ (see Corollary 5.2.5). However, in general we do not have convergence in $\mathbf{H}_\infty$, i.e., we do not have

$$\lim_{N \to \infty} \|k^N C^N (sI - A^N)^{-1} B^N \rho^N - C(sI - A)^{-1} B\|_{H_\infty} = 0;$$

see Proposition 1 and the counterexample on pages 1146–1147 in Kappel and Salamon [58].

5.3   Approximation of the Strongly Stabilizing Solution

In this section, we prove our main convergence result for Riccati equations for strongly stabilizable-detectable systems. The idea will be to transform the LQ problem to an equivalent stable problem as in Theorem 5.1.1. Then, the resulting stable system is approximated, as was done in the previous section. The convergence results of Section 5.2 then will guarantee the convergence of the Riccati solutions. Without proof, we remark that, as in Kappel and Salomon [58], it can be shown that if a sequence of systems $\Sigma(A^N, B^N, C^N)$ converges strongly to $\Sigma(A, B, C)$ and matrices $F^N \in \mathbb{R}^{m(N) \times k(N)}$ are chosen such that the operator sequences $j^N F^N \pi^N \in \mathcal{L}(Z, U)$ and $i^N (F^N)^* \rho^N \in \mathcal{L}(Z, U)$ converge strongly to F and F^*, respectively, then strong convergence of the system $\Sigma(A^N + B^N F^N, B^N, [\, (C^N)^* \;\; (F^N)^* \,]^*)$ to $\Sigma(A + BF, B, [\, C^* \;\; F^* \,]^*)$ holds, as well.

Theorem 5.3.1   *Consider the algebraic Riccati equation (5.3), and assume that*

1. *the system $\Sigma(A, B, C)$ is strongly stabilizable-detectable (i.e., there exist $F \in \mathcal{L}(Z, U)$ and $L \in \mathcal{L}(Y, Z)$ for which the properties C1–C3 in Definition 5.1.2 are satisfied);*

2. *there exists a sequence of systems $\Sigma(A^N, B^N, C^N)$ that is strongly convergent to $\Sigma(A, B, C)$;*

3. *there exists a sequence of matrices $F^N \in \mathcal{L}(\mathbb{R}^{k(N)}, \mathbb{R}^{m(N)})$ such that*

    a. *$j^N F^N \pi^N \to F$ strongly and $i^N (F^N)^* \rho^N \to F^*$ strongly, where F is the operator from part 1;*

    b. *$A^N + B^N F^N$ is a stable matrix;*

    c. *$\Sigma \left(A^N + B^N F^N, B^N, \begin{bmatrix} F^N \\ C^N \end{bmatrix} \right)$ is uniformly output stable;*

    d. *$\Sigma \left(A^N + B^N F^N, B^N, \begin{bmatrix} F^N \\ C^N \end{bmatrix} \right)$ is uniformly input-output stable;*

4. *there exists a sequence of matrices $L^N \in \mathcal{L}(\mathbb{R}^{m(N)}, \mathbb{R}^{p(N)})$ such that*

    a. *$A^N + L^N C^N$ is a stable matrix;*

    b. *$\Sigma(A^N + L^N C^N, [\, B^N \;\; L^N \,], F^N)$ is uniformly input-output stable.*

Then, for every $z \in Z$,

$$Xz = \lim_{N \to \infty} i^N X^N \pi^N z, \tag{5.12}$$

where $X \in \mathcal{L}(Z)$ is the unique, self-adjoint, nonnegative solution of the algebraic Riccati equation (5.3) and $X^N \in \mathcal{L}(\mathbb{R}^{k(N)})$ are the unique, self-adjoint, nonnegative solutions of the sequence of algebraic Riccati equations

$$(A^N)^* X^N + X^N A^N - X^N B^N (B^N)^* X^N + (C^N)^* C^N = 0. \tag{5.13}$$

Moreover, denoting $F_X = -B^* X$ and $F_{X^N}^N = -(B^N)^* X^N$, the approximating closed-loop systems

$$\Sigma \left(A^N + B^N F_{X^N}^N, B^N, \begin{bmatrix} F_{X^N}^N \\ C^N \end{bmatrix} \right)$$

converge strongly to the infinite-dimensional closed-loop system

$$\Sigma\left(A + BF_X, B, \begin{bmatrix} F_X \\ C \end{bmatrix}\right).$$

Proof We apply Theorem 5.1.1 to the approximating sequence of finite-dimensional systems $\Sigma(A^N, B^N, C^N)$. Since $A^N + B^N F^N$ and $A^N + L^N C^N$ are stable matrices, all the assumptions C1–C3 are satisfied and the solutions to (5.13) are given by

$$X^N = (\Psi_F^N)^* \left[I - \mathbb{F}_F^N((\mathbb{F}_F^N)^* \mathbb{F}_F^N)^{-1}(\mathbb{F}_F^N)^*\right] \Psi_F^N, \tag{5.14}$$

where Ψ_F^N and $\mathbb{F}_F^N$ are the output map and input-output map, respectively, of the system

$$\Sigma\left(A^N + B^N F^N, B^N, \begin{bmatrix} F^N \\ C^N \end{bmatrix}, \begin{bmatrix} I \\ 0 \end{bmatrix}\right).$$

We apply Lemma 5.2.4 to obtain

$$\lim_{N\to\infty} (j^N \oplus k^N)\Psi_F^N \pi^N z = \Psi z \text{ in } \mathbf{L}_2(0,\infty; U \times Y), \tag{5.15}$$

$$\lim_{N\to\infty} i^N (\Psi_F^N)^*(\rho^N \oplus \sigma^N)w = \Psi_F^* w \tag{5.16}$$

for all $z \in Z$ and $w \in \mathbf{L}_2(0,\infty; U \times Y)$, where $j^N \oplus k^N$ denotes the direct sum of j^N and k^N defined by $(j^N \oplus k^N)(u^N, y^N) = (j^N u^N, k^N y^N)$.

Similarly, Lemma 5.2.6 yields

$$(j^N \oplus k^N)\mathbb{F}_F^N \rho^N \to \mathbb{F}_F \text{ strongly}, \tag{5.17}$$

$$j^N (\mathbb{F}_F^N)^*(\rho^N \oplus \sigma^N) \to \mathbb{F}_F^* \text{ strongly} \tag{5.18}$$

as $N \to \infty$. Analogously to the definition of $\mathbb{F}_{LF}$ in the proof of Theorem 5.1.1, we define $\mathbb{F}_{LF}^N$ to be the input-output map of the system

$$\Sigma(A^N + L^N C^N, \begin{bmatrix} B^N & -L^N \end{bmatrix}, -F^N, \begin{bmatrix} I & 0 \end{bmatrix}).$$

Assumption 4b of the present theorem tells us that $\|\mathbb{F}_{LF}^N\|$ is uniformly bounded. Now, from assumption 3d, $\sup \|\mathbb{F}_F^N\| < \infty$ and, from (5.8), we have

$$\|((\mathbb{F}_F^N)^* \mathbb{F}_F^N)^{-1}\| \leq \|\mathbb{F}_{LF}^N\|^2.$$

This shows that $\|((\mathbb{F}_F^N)^* \mathbb{F}_F^N)^{-1}\|$ is bounded from above uniformly in N. We also deduce that $\|j^N((\mathbb{F}_F^N)^* \mathbb{F}^N)^{-1}\rho^N\|$ is uniformly bounded from above, the reason being that we use the induced inner products on $\mathbb{R}^{m(N)}$ and so

$$\|((\mathbb{F}_F^N)^* \mathbb{F}_F^N)^{-1}\| = \|j^N((\mathbb{F}_F^N)^* \mathbb{F}^N)^{-1}\rho^N\|.$$

Thus,

$$\begin{aligned}
&\|j^N((\mathbb{F}_F^N)^* \mathbb{F}_F^N)^{-1}\rho^N v - (\mathbb{F}_F^* \mathbb{F}_F)^{-1}v\| \\
&\leq \|j^N((\mathbb{F}_F^N)^* \mathbb{F}_F^N)^{-1}\rho^N v - j^N \rho^N(\mathbb{F}_F^* \mathbb{F}_F)^{-1}v\| \\
&\quad + \|j^N \rho^N(\mathbb{F}_F^* \mathbb{F}_F)^{-1}v - (\mathbb{F}_F^* \mathbb{F}_F)^{-1}v\| \\
&= \|j^N((\mathbb{F}_F^N)^* \mathbb{F}_F^N)^{-1}\rho^N \left\{\mathbb{F}_F^* \mathbb{F}_F - j^N(\mathbb{F}_F^N)^* \mathbb{F}_F^N \rho^N\right\}(\mathbb{F}_F^* \mathbb{F}_F)^{-1}v\| \\
&\quad + \|(j^N \rho^N - I)(\mathbb{F}_F^* \mathbb{F}_F)^{-1}v\| \\
&\to 0
\end{aligned} \tag{5.19}$$

as $N \to \infty$.

The representations of X and X^N ((5.7) and (5.14), respectively), together with the convergence results in (5.15)–(5.19), show that $i^N X^N \pi^N \to X$ strongly as $N \to \infty$. It is an easy consequence of the strong convergence of the stabilizing solutions that the optimal feedback operator $F^N_{X_N}$ and its adjoint converge strongly to their infinite-dimensional counterparts. The convergence of the approximating closed-loop systems now follows from the remark just before Theorem 5.3.1.   □

In the above theorem we have established the strong convergence of the approximating closed-loop systems to the infinite-dimensional closed-loop system. In practice, however, one would like to apply the finite-dimensional feedback to the infinite-dimensional plant. It is easy to see that the resulting closed-loop system

$$\Sigma \left(A + B j^N F^N_{X^N} \pi^N, B, \begin{bmatrix} F^N_{X^N} \\ C \end{bmatrix} \right)$$

does converge strongly to the infinite-dimensional closed-loop system as N tends to infinity. However, we have not been able to show that the resulting system is strongly stable, nor that $A + B j^N F^N_{X^N} \pi^N$ generates a strongly stable semigroup. It probably does not do so, in general.

5.4   Applications to Dissipative Systems with Collocated Actuators and Sensors

We consider linear systems $\Sigma(A, B, B^*)$ under the assumptions A1–A5 that were introduced in Section 2.2. It will turn out that if we choose the approximations of such a system appropriately, i.e., such that the structure is retained in the approximating systems, then most of the assumptions for convergence are automatically satisfied.

The algebraic Riccati equation associated with this system is

$$A^* X z + X A z - X B B^* X z + B B^* z = 0 \tag{5.20}$$

for all $z \in \mathcal{D}(A)$. It follows from Theorem 3.3.2 that there exists a unique, self-adjoint, nonnegative solution X to the algebraic Riccati equation (5.20). We show here that $\|X\| \leq 1$. Indeed,

$$\langle X z_0, z_0 \rangle \leq \mathcal{J}(-B^* z(t), z_0) = 2 \int_0^\infty \|B^* T_B(t) z_0\|^2 dt \leq \|z_0\|^2,$$

which implies that $\|X\| \leq 1$.

In the remainder of this chapter, we assume U and Y to be finite-dimensional, $U = Y = \mathbb{R}^m$. Furthermore, we assume the existence of a sequence of injective linear maps $i^N : \mathbb{R}^{n(N)} \to Z$ and a sequence of surjective linear maps $\pi^N : Z \to \mathbb{R}^{n(N)}$ such that $\pi^N i^N = I_{n(N)}$ and $i^N \pi^N$ is an orthogonal projection on Z. We define the following sequence of approximating systems on $\mathbb{R}^{n(N)}$:

$$\begin{aligned} A^N &= \pi^N A i^N, \\ B^N &= \pi^N B, \\ C^N &= B^* i^N. \end{aligned} \tag{5.21}$$

Lemma 5.4.1 *Let $\Sigma(A^N, B^N, C^N)$ be defined as above. Then, it satisfies $A^N + (A^N)^* \leq 0$ and $C^N = (B^N)^*$. Moreover, $\Sigma(A^N - B^N (B^N)^*, B^N, C^N)$ is uniformly*

*output stable and uniformly input-output stable. If $\Sigma(A^N, B^N, (B^N)^*)$ is observable,
then $A^N - B^N(B^N)^*$ is a stable matrix.*

Proof We first show that $C^N = (B^N)^*$ and $A^N + (A^N)^* \leq 0$. For $u \in \mathbb{R}^m$, $z \in \mathbb{R}^{n(N)}$,

$$
\begin{aligned}
\langle B^N u, z \rangle_{\mathbb{R}^{n(N)}} \\
&= \langle \pi^N B u, z \rangle_{\mathbb{R}^{n(N)}} \\
&= \langle i^N \pi^N B u, i^N z \rangle_Z \\
&= \langle u, B^*(i^N \pi^N)^* i^N z \rangle_{\mathbb{R}^m} \\
&= \langle u, B^* i^N \pi^N i^N z \rangle_{\mathbb{R}^m} \text{ since } i^N \pi^N \text{ is an orthogonal projection} \\
&= \langle u, B^* i^N z \rangle_{\mathbb{R}^m} \text{ since } \pi^N i^N = I_{n(N)} \\
&= \langle u, C^N z \rangle_{\mathbb{R}^m}
\end{aligned}
$$

and so $(B^N)^* = C^N$. Next, consider for $z_1, z_2 \in \mathbb{R}^{n(N)}$,

$$
\begin{aligned}
\langle A^N z_1, z_2 \rangle_{\mathbb{R}^{n(N)}} \\
&= \langle i^N \pi^N A i^N z_1, i^N z_2 \rangle_Z \\
&= \langle A i^N z_1, i^N \pi^N i^N z_2 \rangle_Z \text{ since } i^N \pi^N \text{ is an orthogonal projection} \\
&= \langle A i^N z_1, i^N z_2 \rangle_Z \text{ since } \pi^N i^N = I_{N(n)} \\
&= \langle i^N z_1, A^* i^N z_2 \rangle_Z \\
&= \langle i^N \pi^N i^N z_1, A^* i^N z_2 \rangle_Z \text{ since } \pi^N i^N = I_{N(n)} \\
&= \langle i^N z_1, i^N \pi^N A^* i^N z_2 \rangle_Z \text{ since } i^N \pi^N \text{ is an orthogonal projection} \\
&= \langle z_1, \pi^N A^* i^N z_2 \rangle_{\mathbb{R}^{n(N)}}
\end{aligned}
$$

and so $(A^N)^* = \pi^N A^* i^N$. Now,

$$
\begin{aligned}
\langle (A^N + (A^N)^*) z, z \rangle_{\mathbb{R}^{n(N)}} \\
&= \langle \pi^N (A + A^*) i^N z, z \rangle_{\mathbb{R}^{n(N)}} \\
&= \langle i^N \pi^N (A + A^*) i^N z, i^N z \rangle_Z \\
&= \langle (A + A^*) i^N z, i^N \pi^N i^N z \rangle_Z \text{ since } i^N \pi^N \text{ is an orthogonal projection} \\
&= \langle (A + A^*) i^N z, i^N z \rangle_Z \text{ since } \pi^N i^N = I_{N(n)} \\
&\leq 0 \text{ since } A \text{ is dissipative.}
\end{aligned}
$$

Thus, $A^N + (A^N)^* \leq 0$. Together with $C^N = (B^N)^*$, this shows that for every N,
$\Sigma(A^N - B^N(B^N)^*, B^N, (B^N)^*)$ satisfies properties P4 and P6 of Lemma 2.2.6; hence
it is uniformly output stable and uniformly input-output stable. The stability of $A^N -
B^N(B^N)^*$ follows just as in the infinite-dimensional case from the observability of the pair
$(A^N, (B^N)^*)$ and $A^N + (A^N)^* \leq 0$. Consider solutions $z(t) \in \mathbb{R}^{n(N)}$ of

$$
\dot{z}(t) = (A^N - B^N(B^N)^*)z(t), \quad z(0) = z_0,
$$

for arbitrary $z_0 \in \mathbb{R}^{n(N)}$. We introduce as a Lyapunov function for this differential equation
$V(z) = \|z\|^2$. Then, if we differentiate V along solutions of the differential equation, we
obtain

$$\frac{d}{dt}V(z(t)) = \langle z(t), \dot{z}(t) \rangle + \langle \dot{z}(t), z(t) \rangle$$

$$= \langle z(t), (A^N - B^N(B^N)^*)z(t) \rangle + \langle (A^N - B^N(B^N)^*)z(t), z(t) \rangle$$

$$= \langle (A^N + (A^N)^* - 2B^N(B^N)^*)z(t), z(t) \rangle$$

$$\leq -2\|(B^N)^*z(t)\|^2 \tag{5.22}$$

$$\leq 0,$$

where the inequality follows from the dissipativity of A^N. Now, by Lasalle's invariance principle (see Lasalle [61]), all solutions $z(t)$ converge to the largest $A^N - B^N(B^N)^*$-invariant subset of the set

$$S = \{z \in \mathbb{R}^{n(N)} | \langle (A^N + (A^N)^* - 2B^N(B^N)^*)z, z \rangle = 0\}.$$

Now, let $0 \neq z_0 \in S$; then, for all $t \geq 0$, $z(t) = e^{(A^N - B^N(B^N)^*)t}z_0 \in S$. And so, by (5.22),

$$0 = \frac{d}{dt}V(z(t)) \leq -2\|(B^N)^*z(t)\|^2 = -2\|(B^N)^*e^{(A^N - B^N(B^N)^*)t}z_0\|^2;$$

hence, for all $t \geq 0$, $(B^N)^*e^{(A^N - B^N(B^N)^*)t}z_0 = 0$ which, by the observability assumption, implies that $z_0 = 0$. Thus, $S = \{0\}$ and consequently $A^N - B^N(B^N)^*$ is stable. $\quad\square$

Finally, we give sufficient conditions under which the sequence of finite-dimensional systems $\Sigma(A^N, B^N, (B^N)^*)$, as defined earlier in this section, converges strongly to the infinite-dimensional system $\Sigma(A, B, B^*)$.

Lemma 5.4.2 *Assume that the following conditions are satisfied.*

1. *The sequence of injective maps $i^N : \mathbb{R}^{n(N)} \to Z$ and the sequence of surjective maps $\pi^N : Z \to \mathbb{R}^{n(N)}$ satisfy $\pi^N i^N = I_{n(N)}$, $i^N \pi^N$ is an orthogonal projection on Z and $i^N \pi^N z \to z$ as $N \to \infty$ for all $z \in Z$.*

2. *For all $z \in \mathcal{D}(A)$, there exists a sequence $z^N \in \mathbb{R}^{n(N)}$ such that*

$$\|i^N z^N - z\| \to 0 \text{ and}$$
$$\|i^N A^N z^N - Az\| \to 0 \text{ as } N \to \infty. \tag{5.23}$$

3. *For all $z \in \mathcal{D}(A^*)$, there exists a sequence $z^N \in \mathbb{R}^{n(N)}$ such that*

$$\|i^N z^N - z\| \to 0 \text{ and}$$
$$\|i^N (A^N)^* z^N - A^*z\| \to 0 \text{ as } N \to \infty. \tag{5.24}$$

Then $\Sigma(A^N, B^N, C^N)$ defined by (5.21) converges strongly to $\Sigma(A, B, B^*)$.

Proof From Theorem 3.1.7 of Davies [38, p. 80], we have that (5.23) is equivalent to

$$i^N e^{A^N t} \pi^N z \to T(t)z$$

uniformly on compact time intervals for all $z \in Z$, and (5.24) is equivalent to

$$i^N e^{(A^N)^* t} \pi^N z \to T^*(t)z$$

uniformly on compact time intervals for all $z \in Z$. Next, $i^N B^N = i^N \pi^N B \to B$ as $N \to \infty$ and $C^N \pi^N = (B^N)^* \pi^N = B^* i^N \pi^N \to B^*$ as $N \to \infty$, which completes the proof. $\quad\square$

The following theorem, which is the main result of this section, is now a direct consequence of Theorem 5.3.1, Lemma 5.4.1, Lemma 5.4.2 and the properties of systems $\Sigma(A, B, B^*)$.

Theorem 5.4.3 *Let the system $\Sigma(A, B, B^*)$ satisfy assumptions A1–A5 of Section 2.2 and, furthermore, assume that $U = Y = \mathbb{R}^m$. Define the sequence of approximating systems $\Sigma(A^N, B^N, C^N)$ as in (5.21). Assume that for all N, $\Sigma(A^N, B^N, C^N)$ is observable and, furthermore, that the assumptions 1–3 from Lemma 5.4.2 hold. Then $X^N \to X$ strongly as $N \to \infty$, where X and X^N are the solutions of the Riccati equations associated with $\Sigma(A, B, B^*)$ and $\Sigma(A^N, B^N, C^N)$, respectively.*

Proof By the assumptions 1–3 of Lemma 5.4.2, we have that the sequence of systems $\Sigma(A^N, B^N, C^N)$ as defined in (5.21) converges strongly to $\Sigma(A, B, B^*)$. By Lemma 5.4.1, the sequence of systems $\Sigma(A^N - B^N(B^N)^*, B^N, C^N)$ is uniformly output stable and uniformly input-output stable and, in addition, by the observability assumption these systems are stable. As $\Sigma(A, B, B^*)$ is statically stabilizable with $K = -I$, it is obviously strongly stabilizable-detectable and we can take $F = -B^*$ and $L = -B$. Taking $F^N = -(B^N)^*$ and $L^N = -B^N$, obviously all assumptions of Theorem 5.3.1 are satisfied. Therefore, we can conclude that the stabilizing solutions X^N of the finite-dimensional Riccati equations associated with $\Sigma(A^N, B^N, C^N)$ converge strongly to the strongly stabilizing solution X of the Riccati equation associated with $\Sigma(A, B, B^*)$.   □

In Chapter 9, we describe a model of propagation of sound in a one-dimensional waveguide, terminated by linear oscillators. There we show that it has a state-space realization in our $\Sigma(A, B, B^*)$ class. This allows us to apply the theory of this chapter to that physical example. Therefore, we describe the numerical implementation of an approximation scheme for this example and we present numerical results concerning the approximation of the strongly stabilizing solution of the Riccati equation.

The class of dissipative systems with collocated actuators and sensors has a very special structure. However, one can also allow hybrid systems composed of a distributed system of the structure $\Sigma(A, B, B^*)$ coupled with a finite-dimensional system $\Sigma(A_f, B_f, C_F)$. The composite system

$$\Sigma\left(\begin{bmatrix} A & 0 \\ 0 & A_f \end{bmatrix}, \begin{bmatrix} B \\ B_f \end{bmatrix}, \begin{bmatrix} B^* & C_f \end{bmatrix} \right)$$

will satisfy the assumptions of our theory, provided that the approximating sequence of systems $\Sigma(A^N, B^N, C^N)$ is chosen as before and the finite-dimensional subsystem $\Sigma(A_f, B_f, C_f)$ is minimal. This broadens the class of possible applications. An example of a system with this hybrid structure is the model for a flexible structure with rigid body dynamics in Section 9.4.

Chapter 6

Coprime Factorizations and Dynamic Stabilization

In this chapter, we develop a theory of strong stabilization by dynamic compensators. Because dynamic stabilization is strongly related to the existence of coprime factorizations, we first treat coprime factorizations for strongly stabilizable systems in Section 6.2. In particular, we are interested in finding state-space formulas for coprime factorizations. For both problems, we have not obtained a satisfactory generalization of the finite-dimensional results, and it is not clear that this is possible.

6.1 Problem Formulation

This chapter deals with two important system theoretic problems for strongly stabilizable systems. The first problem is that of finding state-space formulas for coprime factorizations of a given strongly stabilizable system $\Sigma(A, B, C, D)$. Our definition of a coprime factorization is taken from Curtain, Weiss and Weiss [34]. They developed a theory of coprime factorizations for the very general class of regular linear systems. The generality of this class lies in the fact that it allows for very unbounded input and output operators. On the other hand, they considered only exponentially stabilizable systems.

Definition 6.1.1 *Let G be a well-posed transfer function (i.e., G is analytic on some right half-plane). A* left-coprime factorization *of G over $\mathbf{H}_\infty$ is a factorization*

$$G = \tilde{M}^{-1}\tilde{N},$$

where $\tilde{N} \in \mathbf{H}_\infty(\mathcal{L}(U,Y))$, $\tilde{M} \in \mathbf{H}_\infty(\mathcal{L}(Y))$, such that $\tilde{M}(s)$ is invertible on some right half-plane and for which there exist $Y \in \mathbf{H}_\infty(\mathcal{L}(Y,U))$ and $X \in \mathbf{H}_\infty(\mathcal{L}(Y))$ such that for $s \in \mathbb{C}_0^+$,

$$\tilde{M}(s)X(s) - \tilde{N}(s)Y(s) = I. \tag{6.1}$$

This factorization is a normalized *left-coprime factorization if, in addition,*

$$\tilde{M}(j\omega)\tilde{M}^*(j\omega) + \tilde{N}(j\omega)\tilde{N}^*(j\omega) = I. \tag{6.2}$$

A right-coprime factorization *of G over $\mathbf{H}_\infty$ is a factorization*

$$G = NM^{-1},$$

where $N \in \mathbf{H}_\infty(\mathcal{L}(U,Y))$, $M \in \mathbf{H}_\infty(\mathcal{L}(U))$, *such that* $M(s)$ *is invertible on some right half-plane and for which there exist* $\tilde{Y} \in \mathbf{H}_\infty(\mathcal{L}(Y,U))$ *and* $\tilde{X} \in \mathbf{H}_\infty(\mathcal{L}(U))$ *such that for* $s \in \mathbb{C}_0^+$,

$$\tilde{X}(s)M(s) - \tilde{Y}(s)N(s) = I. \tag{6.3}$$

The right-coprime factorization is normalized *if, in addition,*

$$M(j\omega)^* M(j\omega) + N(j\omega)^* N(j\omega) = I. \tag{6.4}$$

A factorization

$$G = NM^{-1} = \tilde{M}^{-1}\tilde{N}$$

is a doubly coprime factorization *of G over* $\mathbf{H}_\infty$ *if* (M,N) *and* $(\tilde{M},\tilde{N})$ *are a right- and left-coprime factorization of G, respectively, and there exist* $X, Y, \tilde{X}, \tilde{Y} \in \mathbf{H}_\infty$ *such that*

$$\begin{bmatrix} M & Y \\ N & X \end{bmatrix}^{-1} = \begin{bmatrix} \tilde{X} & -\tilde{Y} \\ -\tilde{N} & \tilde{M} \end{bmatrix}. \tag{6.5}$$

If, in addition, (6.2) and (6.4) are satisfied, then the factorization is a normalized doubly coprime factorization.

The equations (6.1) and (6.3) are called the Bezout identity, and the functions $X, Y, \tilde{X}, \tilde{Y}$ are the Bezout factors.

The second problem considered in this chapter is that of the construction of dynamic compensators that stabilize a given, strongly stabilizable system $\Sigma(A,B,C,D)$. For this, we consider the feedback interconnection of a plant Σ_G with a compensator Σ_K, as in Figure 6.1. Let Σ_G and Σ_K have transfer function $G(s)$ and $K(s)$, respectively. We have to restrict our consideration to those transfer functions that give rise to a well-posed interconnection.

Definition 6.1.2 *A well-posed transfer function* $K(s)$ *is an* admissible feedback transfer function *for* $G(s)$ *if*

1. $I - G(s)K(s)$ is invertible on some right half-plane, and

2. $(I - G(s)K(s))^{-1}$ is a well-posed transfer function.

In the case that K is an admissible feedback transfer function for G, then $(I - KG)^{-1} = I + K(I - GK)^{-1}G$ also exists and is a well-posed transfer function. The closed-loop transfer function from $\text{col}(v_1,v_2)$ to $\text{col}(y_1,y_2)$ is given by

$$G_{cl} = \begin{bmatrix} (I-GK)^{-1}G & G(I-KG)^{-1}K \\ K(I-GK)^{-1}G & (I-KG)^{-1}K \end{bmatrix}. \tag{6.6}$$

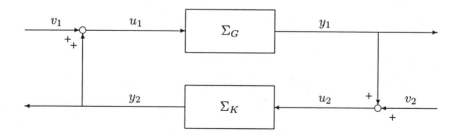

Figure 6.1: Configuration for dynamic compensation.

Often, problems of dynamic stabilization are formulated on the level of transfer functions. In this case, the following notion of input-output stabilization is considered.

Definition 6.1.3 *Let G and K be well-posed transfer functions. K stabilizes G in the* input-output sense *if*

1. *K is an admissible feedback transfer function for G, and*

2. *the closed-loop transfer function from* $\mathrm{col}(v_1, v_2)$ *to* $\mathrm{col}(y_1, y_2)$ *is an element of* $\mathbf{H}_\infty(\mathcal{L}(U \times Y, Y \times U))$.

The input-output stability is often (equivalently) defined in terms of the transfer function from $\mathrm{col}(v_1, v_2)$ to $\mathrm{col}(u_1, u_2)$.

The strong connection between coprime factorizations and dynamic stabilization is well known. First of all, it is known from Smith [95] that a system can be stabilized (in the sense of Definition 6.1.3) if and only if its transfer function has a coprime factorization over $\mathbf{H}_\infty$. A second instance of the relation is the celebrated Youla parameterization of all stabilizing controllers for a given transfer function (see Maciejowski [71] or any other book on multivariable control). We quote two results from Weiss and Curtain [113] about the relation between coprime factorizations over $\mathbf{H}_\infty$ and dynamic stabilization for well-posed transfer functions.

Lemma 6.1.4 *Let G and K be well-posed transfer functions taking values in $\mathcal{L}(U, Y)$ and $\mathcal{L}(Y, U)$, respectively, and suppose that G has a left-coprime factorization $G = \tilde{M}^{-1}\tilde{N}$ and K has a right-coprime factorization $K = UV^{-1}$. Then K stabilizes G in the input-output sense if and only if $\tilde{M}V - \tilde{N}U$ is invertible over $\mathbf{H}_\infty$.*

Theorem 6.1.5 (Youla Parameterization) *Suppose that the well-posed transfer function G has a doubly coprime factorization over $\mathbf{H}_\infty$ as in Definition 6.1.1.*

1. *Assume that K is a well-posed transfer function that stabilizes G. Then K has a right-coprime factorization over $\mathbf{H}_\infty$ if and only if it has a left-coprime factorization over $\mathbf{H}_\infty$. In this case, there exists a unique $Q \in \mathbf{H}_\infty$ such that $X + NQ$ and $\tilde{X} + Q\tilde{N}$ have well-posed inverses and*

$$K = (Y + MQ)(X + NQ)^{-1} = (\tilde{X} + Q\tilde{N})^{-1}(\tilde{Y} + Q\tilde{M}) \qquad (6.7)$$

are right- and left-coprime factorizations of K.

2. *Conversely, if $Q \in \mathbf{H}_\infty$ is such that one of $X + NQ$ or $\tilde{X} + Q\tilde{N}$ has a well-posed inverse, then the other has a well-posed inverse, as well, and the equality in (6.7) holds. In this case, K defined in (6.7) stabilizes G.*

In Section 6.3, we deal with the problem of finding stabilizing dynamic compensators for strongly stabilizable systems, i.e., we try to find systems with transfer function K that stabilize a given strongly stabilizable system with transfer function G in the input-output sense. Furthermore, we study the problem of whether the resulting closed-loop system is actually a strongly stable system. In Chapter 7, we use the theory of this chapter to solve the problem of robust input-output stabilization with respect to coprime factor perturbations.

6.2 Coprime Factorizations

Our aim in this section is to give state-space formulas for doubly coprime factorizations of strongly stabilizable systems. Our treatment of coprime factorizations will stay close to the approach in Curtain, Weiss and Weiss [34]. In that paper, they considered state-space formulas for coprime factorizations of transfer functions of regular linear systems.

For the time being, we consider only coprime factorizations for systems without direct feedthrough (i.e., $D = 0$). The formulas for systems with $D \neq 0$ can be easily constructed from the former ones. The following equations show how coprime factorizations for $H(s) = D + C(sI - A)^{-1}B$ are constructed from a coprime factorization of $G(s) = C(sI - A)^{-1}B$:

$$\begin{bmatrix} M_H & Y_H \\ N_H & X_H \end{bmatrix} = \begin{bmatrix} I & 0 \\ D & I \end{bmatrix} \begin{bmatrix} M_G & Y_G \\ N_G & X_G \end{bmatrix}, \tag{6.8}$$

$$\begin{bmatrix} \tilde{X}_H & -\tilde{Y}_H \\ -\tilde{N}_H & \tilde{M}_H \end{bmatrix} = \begin{bmatrix} \tilde{X}_G & \tilde{Y}_G \\ -\tilde{N}_G & \tilde{M}_G \end{bmatrix} \begin{bmatrix} I & 0 \\ -D & I \end{bmatrix}. \tag{6.9}$$

In the formulas above, the subscript G refers to elements of a coprime factorization of $G(s)$ and the subscript H to those of $H(s)$. These formulas also apply if one wants to construct a coprime factorization of $H(s) = D(s) + G(s)$ in the case that $D(s) \in \mathbf{H}_\infty(\mathcal{L}(U,Y))$.

In the existing theory of coprime factorizations, it is customary to construct coprime factorizations using an exponentially stabilizing feedback operator F and an exponentially detecting injection operator L. In the case of strong stability, this is possible only under extra conditions, which are analogous to those for regular linear systems (see Curtain, Weiss and Weiss [34]).

Theorem 6.2.1 *Assume that the operators F and L are such that the two systems $\Sigma(A + BF, [\ B \ L \ ], [\ F^* \ C^* \ ]^*)$ and $\Sigma(A + LC, [\ B \ L \ ], [\ F^* \ C^* \ ]^*)$ are input-output stable. Then the following constitutes a doubly coprime factorization of $G(s) = C(sI - A)^{-1}B$:*

$$\begin{bmatrix} M_F & Y_F \\ N_F & X_F \end{bmatrix} = \begin{bmatrix} I & 0 \\ 0 & I \end{bmatrix} + \begin{bmatrix} F \\ C \end{bmatrix} (sI - A_F)^{-1} [\ B \ L \ ], \tag{6.10}$$

$$\begin{bmatrix} \tilde{X}_L & -\tilde{Y}_L \\ -\tilde{N}_L & \tilde{M}_L \end{bmatrix} = \begin{bmatrix} I & 0 \\ 0 & I \end{bmatrix} + \begin{bmatrix} F \\ C \end{bmatrix} (sI - A_L)^{-1} [\ -B \ L \ ], \tag{6.11}$$

where $A_F = A + BF$ and $A_L = A + LC$.

Proof It follows directly from the stability assumptions that all components are in $\mathbf{H}_\infty$. That they indeed form a coprime factorization is a matter of algebra, and the computations are the same as in the case of exponential stability (see, for instance, Curtain, Weiss and Weiss [34]).   □

The result above is completely analogous to the corresponding result for regular linear systems in Proposition 3.4 of Curtain, Weiss and Weiss [34]. Their state-space formulas hold under the assumption that $\Sigma(A, B, C)$ is exponentially stabilizable and detectable plus certain regularity conditions on the combination (F, L). These imply the same input-output stability requirements as ours. Staffans has similar results in [98].

Even though our result appears to be nice, there are problems hidden in the assumptions. In contrast to the case of exponentially stabilizable bounded linear systems, in our case it is not at all clear, for a given F and L that are such that $\Sigma(A_F, B, [\ F^*, C^* \ ]^*)$ and $\Sigma(A_L, [\ B \ L \ ], C)$ are both input-output stable, that we also have

$$F(sI - A_F)^{-1}L, \ \ C(sI - A_F)^{-1}L,$$
$$F(sI - A_L)^{-1}B, \ \ F(sI - A_L)^{-1}L$$

in $\mathbf{H}_\infty$. However, we do need to assume this for the candidate Bezout factors X, Y, $\tilde{X}$ and $\tilde{Y}$ to be in $\mathbf{H}_\infty$. Clearly, this problem does not occur in the case of exponentially stabilizable and detectable bounded linear systems. It is exactly this problem that hinders the construction of *normalized* coprime factorizations for strongly stabilizable and detectable systems later in this chapter. It will also hinder the development in Section 6.3, which deals with designing stabilizing compensators for strongly stabilizable systems.

There is one case in which the above mentioned problem does not occur. In the case of a statically stabilizable system, we can choose $F = K_1C$ and $L = BK_2$ where both K_1 and K_2 are statically stabilizing for $\Sigma(A, B, C)$, and this will give us a doubly coprime factorization of $G(s)$.

Next, we give a different construction, which may lead to coprime factorizations for $G(s) = C(sI - A)^{-1}B$. Let $K(s)$ be an admissible feedback transfer function for $G(s)$. We define

$$\begin{bmatrix} M & Y \\ N & X \end{bmatrix} = \begin{bmatrix} (I - KG)^{-1} & K \\ G(I - KG)^{-1} & I \end{bmatrix} \tag{6.12}$$

and

$$\begin{bmatrix} \tilde{X} & -\tilde{Y} \\ -\tilde{N} & \tilde{M} \end{bmatrix} = \begin{bmatrix} I & -K \\ -(I - GK)^{-1}G & (I - GK)^{-1} \end{bmatrix}. \tag{6.13}$$

It is a matter of algebra to check that $G = \tilde{M}^{-1}\tilde{N} = NM^{-1}$ and that

$$\begin{bmatrix} M & Y \\ N & X \end{bmatrix}^{-1} = \begin{bmatrix} \tilde{X} & -\tilde{Y} \\ -\tilde{N} & \tilde{M} \end{bmatrix}.$$

So, (6.12), (6.13) define a coprime factorization of $G(s)$ if all the components are in $\mathbf{H}_\infty$. We give a number of situations for which this is the case.

Lemma 6.2.2 *Let $G(s) = C(sI - A)^{-1}B$ and let $M, N, Y, X, \tilde{M}, \tilde{N}, \tilde{Y}, \tilde{X}$ be defined as in (6.12), (6.13).*

1. *If $K \in \mathbf{H}_\infty$ is such that $(I - GK)^{-1}$ is a well-posed transfer function and $G(I - KG)^{-1} \in \mathbf{H}_\infty$, then (6.12), (6.13) define a doubly coprime factorization of G.*

2. *If $K \in \mathbf{H}_\infty$ stabilizes G in the input-output sense, then (6.12), (6.13) define a doubly coprime factorization of G.*

3. *If $\Sigma(A, B, C)$ is statically stabilizable, i.e., there exists a $K \in \mathcal{L}(Y, U)$ such that $\Sigma(A_K, B, C)$ is a strongly stable system (where $A_K = A + BKC$), then (6.12), (6.13) define a doubly coprime factorization of G. In this case, we have the following state-space formulas:*

$$\begin{bmatrix} M_K & Y_K \\ N_K & X_K \end{bmatrix} = \begin{bmatrix} I + KC(sI - A_K)^{-1}B & K \\ C(sI - A_K)^{-1}B & I \end{bmatrix},$$

$$\begin{bmatrix} \tilde{X}_K & -\tilde{Y}_K \\ -\tilde{N}_K & \tilde{M}_K \end{bmatrix} = \begin{bmatrix} I & -K \\ -C(sI - A_K)^{-1}B & I + C(sI - A_K)^{-1}BK \end{bmatrix}.$$

Proof 1. From the assumptions, it follows that $G(I - KG)^{-1}K \in \mathbf{H}_\infty$. Consequently, $(I - GK)^{-1} \in \mathbf{H}_\infty$, because

$$I + G(I - KG)^{-1}K = I + GK(I - GK)^{-1}$$
$$= (I - GK + GK)(I - GK)^{-1}$$
$$= (I - GK)^{-1}.$$

By assumption, $(I - GK)^{-1}G = G(I - KG)^{-1} \in \mathbf{H}_\infty$ and so $K(I - GK)^{-1}G \in \mathbf{H}_\infty$, which implies via a similar calculation as above that $(I - KG)^{-1} \in \mathbf{H}_\infty$.

2. This is a direct consequence of part 1, because $G(I - KG)^{-1}$ is one of the components of the closed-loop transfer function G_{cl} (see (6.6)).

3. This is a direct consequence of part 2. In this case, the operator K can be viewed as a (constant) transfer function in $\mathbf{H}_\infty$. Because it strongly stabilizes $\Sigma(A, B, C)$, the closed-loop system is certainly input-output stable. The state-space formulas follow by easy calculations from (6.12), (6.13).   □

Note that case 3 is a special instance of case 2, which itself is a special case of 1. Furthermore, note that the factorization for statically stabilizable systems in case 3 is different from the one obtained by taking $F = KC$, $L = BK$ in Theorem 6.2.1.

6.2.1 State-Space Formulas for Normalized Coprime Factorizations

We are especially interested in normalized doubly coprime factorizations because of their important role in the solution of the robust stabilization problem. They are usually constructed with the use of the stabilizing solutions of the control algebraic Riccati equation and the filter algebraic Riccati equation. In the normalized case, we have to take D directly into account, because our earlier procedure (see (6.8) and (6.9)) to incorporate D in a later stage will not retain the normalization property. Let us therefore consider the control and filter Riccati equations for $\Sigma(A, B, C, D)$,

$$A^*Qz_1 + QAz_1 - (QB + C^*D)(I + D^*D)^{-1}(B^*Q + D^*C)z_1 + C^*Cz_1 = 0, \quad (6.14)$$

$$APz_2 + PA^*z_2 - (PC^* + BD^*)(I + DD^*)^{-1}(CP + DB^*)z_2 + BB^*z_2 = 0 \quad (6.15)$$

for $z_1 \in \mathcal{D}(A)$ and $z_2 \in \mathcal{D}(A^*)$. By Theorem 3.3.2 (which can be extended to the $D \neq 0$ case by the same transformation as the one in Section 4.5.4), these equations both have strongly stabilizing solutions if both $\Sigma(A, B, C)$ and $\Sigma(A^*, C^*, B^*)$ are strongly stabilizable and strongly detectable. The corresponding optimal feedback and injection operators are given by

$$F_Q = -(I + D^*D)^{-1}(B^*Q + D^*C), \quad (6.16)$$

$$L_P = -(PC^* + BD^*)(I + DD^*)^{-1}. \quad (6.17)$$

If one takes the same formulas as in Theorem 6.2.1, then, after a normalization, the candidate formulas for the normalized coprime factors would be

$$M_Q(s) = S^{-\frac{1}{2}} + F_Q(sI - A_{F_Q})^{-1}BS^{-\frac{1}{2}}, \quad (6.18)$$

$$N_Q(s) = DS^{-\frac{1}{2}} + (C + DF_Q)(sI - A_{F_Q})^{-1}BS^{-\frac{1}{2}}, \quad (6.19)$$

$$\tilde{M}_P(s) = R^{-\frac{1}{2}} + R^{-\frac{1}{2}}C(sI - A_{L_P})^{-1}L_P, \quad (6.20)$$

$$\tilde{N}_P(s) = R^{-\frac{1}{2}}D + R^{-\frac{1}{2}}C(sI - A_{L_P})^{-1}(B + L_PD), \quad (6.21)$$

where $S = I + D^*D$ and $R = I + DD^*$. Because P and Q are strongly stabilizing solutions to their respective Riccati equations, the functions above are all in $\mathbf{H}_\infty$. However, this is not at all clear for the candidate Bezout factors

$$\tilde{X}(s) = R^{\frac{1}{2}} + R^{\frac{1}{2}}F_Q(sI - A_{L_P})^{-1}(B + L_PD), \quad (6.22)$$

$$\tilde{Y}(s) = R^{\frac{1}{2}}F_Q(sI - A_{L_P})^{-1}L_P, \quad (6.23)$$

$$X(s) = S^{\frac{1}{2}} + (C + DF_Q)(sI - A_{F_Q})^{-1}L_PS^{\frac{1}{2}}, \quad (6.24)$$

$$Y(s) = F_Q(sI - A_{F_Q})^{-1}L_PS^{\frac{1}{2}}. \quad (6.25)$$

The problem is the one described below Theorem 6.2.1: we have been unable to show that the terms containing both F_Q and L_P are in $\mathbf{H}_\infty$. Because P and Q are strongly stabilizing solutions, it does follow, however, that for all $u \in U$ and $y \in Y$, the functions $\tilde{Y}(s)y, Y(s)u$, $(\tilde{X}(s) - R^{\frac{1}{2}})u$ and $(X(s) - S^{\frac{1}{2}})y$ are in $\mathbf{H}_2$.

Despite this problem, we have been able to show that (6.18)–(6.21) constitute a normalized doubly coprime factorization for strongly stabilizable and detectable systems with finite-dimensional inputs and outputs under an additional assumption on the spectrum of A. The approach is based on a matrix generalization of the *Carleson Corona Theorem*. Unfortunately, this approach does not provide us with formulas for the Bezout factors, even though they must exist. Also, the proof does not give any information about whether the functions in (6.22)–(6.25) are in $\mathbf{H}_\infty$. This remains an interesting open question.

At the end of this section, we show that for systems in the $\Sigma(A, B, B^*)$ class with nonnegative, feedthrough operator D, we don't need the assumption of finite-dimensional input and output spaces. For this particular case, we have obtained state-space formulas for Bezout factors.

As mentioned, the approach to show that (M_Q, N_Q) and $(\tilde{M}_P, \tilde{N}_P)$ are coprime is based on a matrix generalization of the Carleson Corona Theorem (see Theorem 14-10 in Fuhrmann [46, p. 204] or Theorem 3.1 in [45]). For easy reference, we abstract a special case of the results.

Theorem 6.2.3

1. Given $M \in \mathbf{H}_\infty(\mathbb{C}^{m \times m})$ and $N \in \mathbf{H}_\infty(\mathbb{C}^{k \times m})$, a necessary and sufficient condition for M and N to be coprime over $\mathbf{H}_\infty$ is the existence of a $\delta > 0$ such that

$$\inf_{s \in \mathbb{C}_0^+} \left\{ \|M(s)u\|^2 + \|N(s)u\|^2, \ u \in \mathbb{C}^m, \ \|u\| = 1 \right\} \geq \delta, \qquad (6.26)$$

where $\|\cdot\|$ denotes the Euclidean norm.

2. Given $\tilde{M} \in \mathbf{H}_\infty(\mathbb{C}^{k \times k})$ and $\tilde{N} \in \mathbf{H}_\infty(\mathbb{C}^{k \times m})$, a necessary and sufficient condition for $\tilde{M}$ and $\tilde{N}$ to be coprime over $\mathbf{H}_\infty$ is the existence of a $\delta > 0$ such that

$$\inf_{s \in \mathbb{C}_0^+} \left\{ \|\tilde{M}(s)^* y\|^2 + \|\tilde{N}(s)^* y\|^2, \ y \in \mathbb{C}^k, \ \|y\| = 1 \right\} \geq \delta. \qquad (6.27)$$

We remark that in (6.26), (6.27), Fuhrmann used $\|M(s)u\| + \|N(s)u\|$ instead of $\|M(s)u\|^2 + \|N(s)u\|^2$. Of course, they are equivalent statements, but the version given above is more convenient for our purposes. Using this result, we can give sufficient conditions for the functions in (6.18)–(6.21) to constitute a normalized doubly coprime factorization.

Theorem 6.2.4 *Suppose that $\Sigma(A, B, C, D)$ is such that A generates a C_0-semigroup on the Hilbert space Z, $B \in \mathcal{L}(\mathbb{C}^m, Z)$, $C \in \mathcal{L}(Z, \mathbb{C}^k)$, $D \in \mathcal{L}(\mathbb{C}^m, \mathbb{C}^k)$.*

1. Under the following assumptions, (M_Q, N_Q) given by (6.18)–(6.19) constitutes a normalized right-coprime factorization of $G = D + C(sI - A)^{-1}B$:

(i) *A has no essential spectrum on the imaginary axis.*

(ii) *There exists an $F \in \mathcal{L}(Z, \mathbb{C}^m)$ such that $\begin{bmatrix} F^* & C^* \end{bmatrix}^* (sI - A_F)^{-1} z \in \mathbf{H}_2(U \times Y)$ for all $z \in Z$.*

(iii) *$\Sigma(A, B, C, D)$ is strongly detectable.*

2. Under the following assumptions, $(\tilde{M}_P, \tilde{N}_P)$ given by (6.20)–(6.21) constitutes a normalized left-coprime factorization of $G = D + C(sI - A)^{-1}B$:

(iv) A^* has no essential spectrum on the imaginary axis.

(v) There exists an $L \in \mathcal{L}(\mathbb{C}^k, Z)$ such that $\begin{bmatrix} B & L \end{bmatrix}^* (sI - A_L^*)^{-1} z \in \mathbf{H}_2(U \times Y)$ for all $z \in Z$.

(vi) $\Sigma(A^*, C^*, B^*, D^*)$ is strongly detectable.

Proof We prove only part 1, as part 2 is its dual.

(a) By conditions (ii) and (iii), the control Riccati equation (6.14) has a strongly stabilizing solution which is the unique, self-adjoint, nonnegative solution (this follows from Theorem 3.3.2; see also Remark 3.3.3). Since A_{F_Q} is a finite-rank perturbation of A, their essential spectrum is identical (see Theorem 5.35 in Kato [59, p. 244]) and so the essential spectrum of A_{F_Q} is not on the imaginary axis. Together with the strong stability of the semigroup generated by A_{F_Q}, this implies that A_{F_Q} has no spectrum in $\mathrm{Re}(s) \geq 0$.

(b) We now show that $\tilde{M}_Q$ and $\tilde{N}_Q$ satisfy (6.26). By direct calculation, following Curtain and Zwart [36, p. 372], we find that

$$
S^{\frac{1}{2}}(M_Q(s)^* M_Q(s) + N_Q(s)^* N_Q(s))S^{\frac{1}{2}}
$$
$$
= S + B^*(\bar{s}I - A_{F_Q}^*)^{-1}(F_Q^* F_Q + C_Q^* C_Q)(sI - A_{F_Q})^{-1}B
$$
$$
+ (F_Q + D^* C_Q)(sI - A_{F_Q})^{-1}B + B^*(\bar{s}I - A_{F_Q}^*)^{-1}(F_Q^* + C_Q^* D),
$$

where $C_Q = C + DF_Q$. Using the Riccati equation (6.14) and the identities $F_Q + D^* C_Q = -B^* Q$, $F_Q^* F_Q + C_Q^* C_Q = \tilde{C}^* \tilde{C} + Q\tilde{B}\tilde{B}^* Q$ from Curtain and Zwart [36, p. 372] (where $\tilde{C} = R^{-\frac{1}{2}}C$ and $\tilde{B} = BS^{-\frac{1}{2}}$), the above simplifies to

$$
M_Q(s)^* M_Q(s) + N_Q(s)^* N_Q(s) = I - 2\mathrm{Re}(s)\tilde{B}^*(\bar{s}I - A_{F_Q}^*)^{-1}Q(sI - A_{F_Q})^{-1}\tilde{B},
$$

and so

$$
\|M_Q(s)u\|^2 + \|N_Q(s)u\|^2 = \|u\|^2 - 2\mathrm{Re}(s)\|Q^{\frac{1}{2}}(sI - A_{F_Q})^{-1}\tilde{B}u\|^2, \quad (6.28)
$$

which holds for $\mathrm{Re}(s) \geq 0$ since from part (a) of the proof, $\sigma(A_{F_Q}) \cap \overline{\mathbb{C}_0^+} = \emptyset$. In particular, there holds

$$
\|M_Q(i\omega)u\|^2 + \|N_Q(i\omega)u\|^2 = \|u\|^2 \quad \text{for all } \omega \in \mathbb{R}, \quad (6.29)
$$

whence $\left\| \begin{bmatrix} M_Q \\ N_Q \end{bmatrix} \right\|_\infty = 1$.

Recall that the candidate Bezout factors $\tilde{X}$ and $\tilde{Y}$ in (6.22), (6.23) are in $\mathbf{H}_2$ (modulo a constant) and so the shifted factors $\tilde{X}(s - \varepsilon)$, $\tilde{Y}(s - \varepsilon)$ are in $\mathbf{H}_\infty$ for each $\varepsilon > 0$. In effect, we have a coprime factorization on $\mathrm{Re}(s) > \varepsilon$,

$$
\tilde{X} M_Q - \tilde{Y} N_Q = I
$$

with M_Q, N_Q, $\tilde{X}$, $\tilde{Y}$ all holomorphic and bounded on $\mathrm{Re}(s) > \varepsilon$. Applying Theorem 6.2.3 to this situation, we conclude that there exists a $\delta(\varepsilon) > 0$ such that

$$
\inf_{s \in \mathbb{C}_\varepsilon^+} \{\|M_Q(s)u\|^2 + \|N_Q(s)u\|^2, \ u \in \mathbb{C}^m, \ \|u\| = 1\} \geq \delta(\varepsilon). \quad (6.30)
$$

We need to show that $\delta(\varepsilon)$ has a nonzero lower limit as ε tends to zero. Suppose, on the contrary, that there exists a sequence $u_n \in \mathbb{C}^m$, $\|u_n\| = 1$ and $z_n \in \bar{\mathbb{C}}_0^+$ such that $\|M_Q(z_n)u_n\| < 1/n$ and $\|N_Q(z_n)u_n\| < 1/n$. Since u_n is a bounded sequence in $\mathbb{C}^m$, it has a convergent subsequence $u_{n_k} \to \xi$ as $k \to \infty$, and we have

$$
\begin{aligned}
\|M_Q(z_{n_k})\xi\| &\leq \|M_Q(z_{n_k})(\xi - u_{n_k})\| + \|M_Q(z_{n_k})u_{n_k}\| \\
&\leq \|\xi - u_{n_k}\| + 1/n_k \quad \text{since } \|M_Q\|_\infty \leq 1 \\
&< 2/n_k
\end{aligned}
$$

for sufficiently large k. Similarly, $\|N_Q(z_{n_k})\xi\| < 2/n_k$. Now, (6.30) with $u = \xi$ implies that for sufficiently large n_k, $\mathrm{Re}(z_{n_k}) \leq \varepsilon$. Since ε can be chosen arbitrarily small and (6.29) with $u = \xi$ holds on the imaginary axis, we can conclude that there exists no such sequence z_n in $\mathbb{C}_0^+$.

(c) So, we have shown that (M_Q, N_Q) represents a normalized right-coprime factorization for G. $\quad\Box$

We end this section by giving two important corollaries to this theorem.

Corollary 6.2.5 *Consider a system $\Sigma(A, B, C, D)$ with finite-dimensional input and output spaces. Under the conditions (i)–(vi) (see Theorem 6.2.4), this system has a normalized doubly coprime factorization $G = N_Q M_Q^{-1} = \tilde{M}_P^{-1} \tilde{N}_P$, with coprime factors given by (6.18)–(6.21).*

Proof This follows directly from Theorem 6.2.4 above and Lemma A.7.44 in Curtain and Zwart [36]. $\quad\Box$

A more balanced, slightly stronger version of Theorem 6.2.4 is the following corollary.

Corollary 6.2.6 *Suppose that $\Sigma(A, B, C, D)$ is such that A generates a C_0-semigroup on the Hilbert space Z, $B \in \mathcal{L}(\mathbb{C}^m, Z)$, $C \in \mathcal{L}(Z, \mathbb{C}^k)$ and $D \in \mathcal{L}(\mathbb{C}^m, \mathbb{C}^k)$. If A has compact resolvent and $\Sigma(A, B, C, D)$ is strongly stabilizable and strongly detectable, then (M_Q, N_Q) and $(\tilde{M}_P, \tilde{N}_P)$ given by (6.18)–(6.21) are normalized right-coprime and left-coprime factorizations of $G(s)$, respectively.*

Proof If A has compact resolvent, then (i) and (iv) hold and $A + BF$ generates a strongly stable semigroup if and only if $A^* + F^*B^*$ does. $\quad\Box$

A consequence of the above theorem is that there exist Bezout factors $X, Y, \tilde{X}, \tilde{Y} \in \mathbf{H}_\infty$ satisfying (6.5). Note that Bezout factors are not unique and we have not shown that the candidate Bezout factors $\tilde{X}, \tilde{Y}, X, Y$ in (6.22)–(6.25) are true Bezout factors. We know only that they are in $\mathbf{H}_2$ (modulo a constant), but not that they are in $\mathbf{H}_\infty$. While this is not essential for the robust stabilization problem with respect to coprime factor perturbations (which we address in Chapter 7), it is needed for dynamic stabilization results, which are the topic of the next section. This remains an interesting open question.

It is interesting to note that the example in Section 9.4, a model describing the dynamics of a flexible structure, does satisfy the assumptions of Theorem 6.2.4. Hence, it has a normalized doubly coprime factorization given by (6.18)–(6.21).

Another important class of systems satisfying the conditions of Corollary 6.2.6 is that of systems in the $\Sigma(A, B, B^*)$ class with finite-dimensional input space U.

We can relieve the assumption on the dimension of U for systems in the $\Sigma(A, B, B^*)$ class with a nonnegative feedthrough operator D. The remainder of this section will be devoted to this subject. The approach is to construct explicitly Bezout factors for this class. We need the following result, which is taken from Curtain and van Keulen [33].

Lemma 6.2.7 *If $G(s)$ is positive real, then*

$$M(s) = \tilde{M}(s) = (I + G(s))^{-1},$$
$$N(s) = \tilde{N}(s) = G(s)(I + G(s))^{-1},$$
$$X(s) = \tilde{X}(s) = -Y(s) = -\tilde{Y}(s) = I$$

form a doubly coprime factorization of G.

The following result is very close to Theorem 12 in Curtain and Zwart [37]. They needed to restrict their theorem to a finite-dimensional input space.

Theorem 6.2.8 *Consider the system $\Sigma(A, B, B^*, D)$ under assumptions A1–A2 (see page 17). Assume, furthermore, that $D = D^* \geq 0$. Then the following define a normalized doubly coprime factorization of $G(s) = D + B^*(sI - A)^{-1}B$:*

$$M_Q = S^{-\frac{1}{2}} + F_Q(sI - A_{F_Q})^{-1}BS^{-\frac{1}{2}},$$
$$N_Q = DS^{-\frac{1}{2}} + (B^* + DF_Q)(sI - A_{F_Q})^{-1}BS^{-\frac{1}{2}},$$
$$\tilde{M}_P = S^{-\frac{1}{2}} + S^{-\frac{1}{2}}B^*(sI - A_{L_P})^{-1}L_P,$$
$$\tilde{N}_P = S^{-\frac{1}{2}}D + S^{-\frac{1}{2}}B^*(sI - A_{L_P})^{-1}(B + L_P D),$$
$$\tilde{X}(s) = -\tilde{Y}(s) = (N_Q(s) + M_Q(s))^{-1},$$
$$X(s) = -Y(s) = (\tilde{N}_P(s) + \tilde{M}_P(s))^{-1},$$

where $S = I + D^2$, $F_Q = -S^{-1}(B^*Q + DB^*)$, $L_P = -(PB + BD)S^{-1}$, $A_{F_Q} = A + BF_Q$, $A_{L_P} = A + L_P B^*$. P and Q are the minimal self-adjoint, nonnegative solutions of the Riccati equations

$$A^*Qz_1 + QAz_1 - (QB + BD)S^{-1}(B^*Q + D^*B^*)z_1 + BB^*z_1 = 0,$$
$$APz_2 + PA^*z_2 - (PB + BD^*)S^{-1}(B^*P + DB^*)z_2 + BB^*z_2 = 0$$

for $z_1 \in \mathcal{D}(A)$ and $z_2 \in \mathcal{D}(A^*)$.

Proof Our assumptions ensure that $\Sigma(A, B, B^*, D)$ satisfies properties P4 and P5 from Lemma 2.2.6. Thus, we can apply Theorem 3.3.1 (see also Remark 3.3.3) to show that each Riccati equation has a minimal self-adjoint, nonnegative solution, Q and P, respectively. Moreover, Theorem 3.3.1 shows that F_Q and L_P are such that their corresponding closed-loop systems are input-output stable. Consequently, $\tilde{M}_P, \tilde{N}_P, M_Q$ and $N_Q \in \mathbf{H}_\infty$. It is a matter of long but straightforward computation to show that $M_Q, N_Q, \tilde{M}_P, \tilde{N}_P, \tilde{Y}, \tilde{X}, Y, X$ satisfy (6.1)–(6.5). For the normalization properties (6.2) and (6.4), these computations are the same as the corresponding ones in Theorem 7.3.11 and Exercise 7.29 in Curtain and Zwart [36] which make use of the fact that P and Q satisfy their respective Riccati equations.

To show that the given factorization is indeed a *coprime* factorization, we need to show that $X, Y, \tilde{X}, \tilde{Y} \in \mathbf{H}_\infty(\mathcal{L}(U))$. We will do this only for $\tilde{X}(s) = (N_Q + M_Q)^{-1}$, as the proof for $X(s)$ is analogous. Now,

$$M_Q(s) + N_Q(s) = (I + D)S^{-\frac{1}{2}} + (B^* + (I + D)F_Q)(sI - A_{F_Q})^{-1}BS^{-\frac{1}{2}}$$
$$= (I + D)\left\{I + (F_Q + (I + D)^{-1}B^*)(sI - A_{F_Q})^{-1}B\right\}S^{-\frac{1}{2}},$$

so,

$$S^{-\frac{1}{2}}\tilde{X}(s)(I + D) = S^{-\frac{1}{2}}(M(s) + N(s))^{-1}(I + D)$$
$$= I - (F_Q + (I + D)^{-1}B^*)(sI - A_{F_Q} + BF_Q + B(I + D)^{-1}B^*)^{-1}B$$
$$= I - (F_Q + (I + D)^{-1}B^*)(sI - A + B(I + D)^{-1}B^*)^{-1}B.$$

Because $D \geq 0$ implies that $(I+D)^{-1} > \varepsilon I$ for some $\varepsilon > 0$, we can factor $(I+D)^{-1} = LL^*$, with L boundedly invertible. Now, the system $\Sigma(A, BL, L^*B^*)$ satisfies assumptions A1 and A2, because $\Sigma(A, B, B^*)$ does so. Hence, property P5 of Lemma 2.2.6 applies, and it follows that $\Sigma(A - BLL^*B^*, BL, L^*B^*)$ is input stable, i.e.,

$$L^*B^*(sI - A_{BL}^*)^{-1}z \in \mathbf{H}_2(U),$$

where $A_{BL} = A - BLL^*B = A - B(I+D)^{-1}B^*$. By the invertibility of L, we obtain $B^*(sI - A^* + B(I+D)^{-1}B^*)^{-1}z \in \mathbf{H}_2(U)$ for all $z \in Z$. Consequently,

$$(\tilde{X}^{\sim}(-s) - S^{\frac{1}{2}})u \in \mathbf{H}_2(U). \tag{6.31}$$

Next, we show that $\tilde{X}(j\omega) \in \mathbf{L}_\infty(-j\infty, j\infty; \mathcal{L}(U))$. To do so, note that

$$\tilde{X}(s) = (N_Q(s) + M_Q(s))^{-1} = M_Q^{-1}(s)(I + G(s))^{-1}.$$

Therefore, $M_Q(s)\tilde{X}(s) = (I + G(s))^{-1}$ and $N_Q(s)\tilde{X}(s) = G(s)(I + G(s))^{-1}$. Because $G(s)$ is positive real, we can apply Lemma 6.2.7 to show that $M_Q(s)\tilde{X}(s)$ and $N_Q(s)\tilde{X}(s) \in \mathbf{H}_\infty(\mathcal{L}(U))$. From the normalization of the factorization, we obtain

$$\tilde{X}^*(j\omega)\tilde{X}(j\omega) = \tilde{X}^*(j\omega)N_Q^*(j\omega)N_Q(j\omega)\tilde{X}(j\omega) + \tilde{X}^*(j\omega)M_Q^*(j\omega)M_Q(j\omega)\tilde{X}(j\omega).$$

By the above observation, the right-hand side is in $\mathbf{L}_\infty(-j\infty, j\infty; \mathcal{L}(U))$ and so the same holds for the left-hand side. Thus, $\tilde{X}^{\sim}(j\omega) \in \mathbf{L}_\infty(-j\infty, j\infty; \mathcal{L}(U))$. Combining this with (6.31), we obtain that $\tilde{X}(s) \in \mathbf{H}_\infty(\mathcal{L}(U))$, by Lemma 4.2.4. $\square$

6.3 Dynamic Stabilization

In this section, we investigate the feasibility of the use of observer-based dynamic compensators for stabilization of strongly stabilizable systems. We restrict ourselves to systems without direct feedthrough (i.e., the case $D = 0$). The extension to the case that $D \neq 0$ is straightforward and comes only at the cost of more complicated formulas. The setup that we consider is the classical one and it is schematically represented in Figure 6.2.

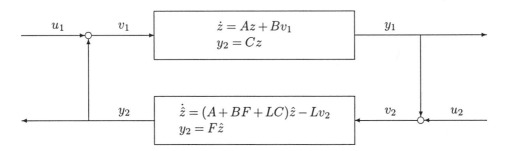

Figure 6.2: Configuration for observer-based control.

Writing down the dynamics in terms of the plant-state z and the state-estimation error $e := z - \hat{z}$, we obtain the equations

$$\begin{pmatrix} \dot{z} \\ \dot{e} \end{pmatrix} = \begin{bmatrix} A + BF & -BF \\ 0 & A + LC \end{bmatrix} \begin{pmatrix} z \\ e \end{pmatrix} + \begin{bmatrix} B & 0 \\ B & L \end{bmatrix} \begin{pmatrix} u_1 \\ u_2 \end{pmatrix},$$

$$\begin{pmatrix} y_1 \\ y_2 \end{pmatrix} = \begin{bmatrix} C & 0 \\ F & -F \end{bmatrix} \begin{pmatrix} z \\ e \end{pmatrix}.$$

In what folllows, we denote $A_F = A + BF$ and $A_L = A + LC$. Mimicking the case of exponentially stabilizable systems, we take F and L to be a strongly stabilizing feedback operator and a strongly detecting injection operator, respectively. More precisely, we assume that

S1. $F \in \mathcal{L}(Z, U)$ is such that $\Sigma\left(A_F, B, \begin{bmatrix} F \\ C \end{bmatrix}\right)$ is a strongly stable system, and

S2. $L \in \mathcal{L}(Y, Z)$ is such that $\Sigma(A_L, [\, B \;\; L \,], C)$ is a strongly stable system.

We stress that assumptions S1 and S2 are stronger than just strong stabilizability or strong detectability. The difference is that in S1 we also demand input-output stability and input stability and that in S2 we also demand input-output stability and output stability.

The strong connection between coprime factorizations and dynamic stabilization is well known and was described in Section 6.1. Indeed, problems similar to the ones that were found in the previous section arise here.

We start by turning our attention to the question of what compensators of the form sketched above do achieve input-output stability of the closed-loop system. Later in this section, we consider strong stability of the closed-loop system. The closed-loop transfer function is given by

$$G_{cl}(s) = \begin{bmatrix} G_{11} & G_{12} \\ G_{21} & G_{22} \end{bmatrix},$$

where

$$\begin{aligned} G_{11} &= C(sI - A_F)^{-1}B(I - F(sI - A_L)^{-1}B), \\ G_{12} &= -C(sI - A_F)^{-1}BF(sI - A_L)^{-1}L, \\ G_{21} &= -I + (I + F(sI - A_F)^{-1}B)(I - F(sI - A_L)^{-1}B), \\ G_{22} &= -(I + F(sI - A_F)^{-1}B)F(sI - A_L)^{-1}L. \end{aligned} \tag{6.32}$$

If we describe the closed-loop dynamics in terms of $\hat{z}$ and e instead of z and e, we obtain a different representation of this transfer function:

$$\begin{aligned} G_{11} &= (I - C(sI - A_F)^{-1}L)C(sI - A_L)^{-1}B, \\ G_{12} &= -I + (I - C(sI - A_F)^{-1}L)(I + C(sI - A_L)^{-1}L), \\ G_{21} &= -F(sI - A_F)^{-1}LC(sI - A_L)^{-1}B, \\ G_{22} &= -F(sI - A_F)^{-1}L(I + C(sI - A_L)^{-1}L). \end{aligned} \tag{6.33}$$

We now have the following theorem.

Theorem 6.3.1 *Let $F \in \mathcal{L}(Z,U)$ and $L \in \mathcal{L}(Y,Z)$. Consider the conditions*

S3. $\Sigma\left(A_L, [\; B \;\; L \;], \begin{bmatrix} F \\ C \end{bmatrix}\right)$ *is input-output stable, and*

S4. $\Sigma\left(A_F, [\; B \;\; L \;], \begin{bmatrix} F \\ C \end{bmatrix}\right)$ *is input-output stable.*

If either S1 and S3 or S2 and S4 hold, then the compensator

$$\begin{aligned} \hat{z} &= (A + BF + LC)\hat{z} - Ly, \\ u &= F\hat{z} \end{aligned} \tag{6.34}$$

stabilizes $\Sigma(A,B,C)$ in the input-output sense.

Proof We show that the combination S1 and S3 is a sufficient condition for stabilization by appealing to (6.32). That S2 and S4 are sufficient as well follows completely analogously by using the alternative representation (6.33) of the closed-loop transfer function.

By S1, the transfer functions $C(sI - A_F)^{-1}B$, $F(sI - A_F)^{-1}B$ are in $\mathbf{H}_\infty$ and by S3, $F(sI - A_L)^{-1}B \in \mathbf{H}_\infty$ and $F(sI - A_L)^{-1}L \in \mathbf{H}_\infty$. This implies that all components of G_{cl} are in $\mathbf{H}_\infty$, and so the closed-loop system is input-output stable. $\qquad\square$

It follows from the proof of the above theorem that S1 and S2 are unnecessarily strong assumptions for this result. It would be sufficient to demand that the systems in S1 and S2 are input-output stable.

Next, we turn our attention to conditions under which application of a compensator does lead to a strongly stable closed-loop system. Clearly, the assumptions of Theorem 6.3.1 will be necessary here because input-output stability is weaker than strong stability. In fact, the conditions that we need in this case are similar to the conditions in the previous theorem.

Theorem 6.3.2 *Let $F \in \mathcal{L}(Z,U)$ and $L \in \mathcal{L}(Y,Z)$. Consider the conditions*

S5. $\Sigma\left(A_L, [\; B \;\; L \;], \begin{bmatrix} F \\ C \end{bmatrix}\right)$ *is a strongly stable system, and*

S6. $\Sigma\left(A_F, [\; B \;\; L \;], \begin{bmatrix} F \\ C \end{bmatrix}\right)$ *is a strongly stable system.*

If either S1 and S5 or S2 and S6 hold, then the compensator (6.34) makes the closed-loop configuration of Figure 6.2 a strongly stable system.

Proof We show that the combination of S1 and S5 is sufficient for strong stabilization by appealing to (6.32). The proof that S2 and S6 are sufficient as well is completely analogous but uses (6.33) instead.

First, we prove the strong stability of the closed-loop semigroup. That is, we show that $(z(t), e(t))$ converge to zero as t tends to infinity. These variables satisfy

$$\begin{aligned} e(t) &= T_L(t)e_0, \\ z(t) &= T_F(t)z_0 - \int_0^t T_F(t-s)BFT_L(s)e_0\,ds, \end{aligned}$$

where $T_L(t)$ and $T_F(t)$ are the semigroups generated by $A + LC$ and $A + BF$, respectively. By S5, $T_L(t)$ is a strongly stable semigroup and so $e(t) \to 0$ as $t \to \infty$. To show that $z(t) \to 0$ as $t \to \infty$, we want to apply Lemma 2.1.3. To do so, we need to show that $T_F(t)$ is a strongly stable semigroup, $\Sigma(A_F, B, -)$ is input stable and that $FT_L(\cdot)e_0 \in \mathbf{L}_2(0, \infty; U)$ for all $e_0 \in Z$. The first two demands are satisfied by S1; the last one is equivalent to the output stability of $\Sigma(A_L, -, F)$, which is guaranteed by assumption S5.

As S5 implies S3, Theorem 6.3.1 applies to show that the closed-loop system is input-output stable.

For output stability of the closed-loop system, we need to show that the following transfer function is in $\mathbf{H}_2(U \times Y)$ for all $z_1, z_2 \in Z$:

$$\begin{bmatrix} F(sI - A_F)^{-1} & -(I + F(sI - A_F)^{-1}B)F(sI - A_L)^{-1} \\ C(sI - A_F)^{-1} & -C(sI - A_F)^{-1}BF(sI - A_L)^{-1} \end{bmatrix} \begin{pmatrix} z_1 \\ z_2 \end{pmatrix}.$$

Sufficient for this to hold are that $C(sI - A_F)^{-1}B \in \mathbf{H}_\infty(\mathcal{L}(U, Y))$, $F(sI - A_F)^{-1}B \in \mathbf{H}_\infty(\mathcal{L}(U))$, $C(sI - A_F)^{-1}z \in \mathbf{H}_2(Y)$, $F(sI - A_F)^{-1}z \in \mathbf{H}_2(U)$ and $F(s - A_L)^{-1}z \in \mathbf{H}_2(U)$ for all $z \in Z$. The last condition follows from S5 and the others from S1.

Finally, input stability of the closed-loop system means that for all $z_1, z_2 \in Z$,

$$\begin{bmatrix} (I - B^*(sI - A_L^*)^{-1}F^*)B^*(sI - A_F^*)^{-1} & B^*(sI - A_L^*)^{-1} \\ -L^*(sI - A_L^*)^{-1}F^*B^*(sI - A_F^*)^{-1} & L^*(sI - A_L^*)^{-1} \end{bmatrix} \begin{pmatrix} z_1 \\ z_2 \end{pmatrix}$$

is in $\mathbf{H}_2(U \times Y)$. This is the case if $B^*(sI - A_F^*)^{-1}z \in \mathbf{H}_2(U)$, $B^*(sI - A_L^*)^{-1}z \in \mathbf{H}_2(U)$, $L^*(sI - A_L^*)^{-1}z \in \mathbf{H}_2(U)$, $F(sI - A_L)^{-1}B \in \mathbf{H}_\infty(\mathcal{L}(U))$ and $F(sI - A_L)^{-1}L \in \mathbf{H}_\infty(\mathcal{L}(U))$. The first condition is satisfied by S1, the others by S5.   □

As in the case of the formulas for coprime factorizations in the previous section, the conditions that we obtained are somewhat problematic. For given F and L satisfying S1 and S2, it is in general not clear whether they also satisfy one of the conditions S3–S6. In particular, it is not clear whether any of these conditions are satisfied for the LQG feedback and injection operators F_Q and L_P. This case suffers from exactly the same problem as the result on normalized doubly coprime factorizations: It is not clear if $F_Q(sI - A_{F_Q})^{-1}L_P$, $C(sI - A_{F_Q})^{-1}L_P$, $F_Q(sI - A_{L_P})^{-1}B$ and $F_Q(sI - A_{L_P})^{-1}L_P$ are in $\mathbf{H}_\infty$. Note that the situation here is even somewhat worse than in Section 6.2. We would need to know that $A + L_PC$ generates a strongly stable semigroup, but the Riccati theory only allows us to conclude that its adjoint generates a strongly stable semigroup. These problems make the present results not completely satisfactory. For statically stabilizable systems, we give a number of special cases in which the conditions are satisfied.

- Take $F = K_1C$ and $L = BK_2$, where for $i = 1, 2$, $\Sigma(A + BK_iC, B, C)$ is a strongly stable system. Then all of the above conditions S1–S6 are satisfied and so the corresponding compensator yields a strongly stable closed-loop system.

- Let L be any strongly detecting injection (i.e., such that S2 is satisfied) and choose $F = KC$, where K is statically stabilizing $\Sigma(A, B, C)$. Then the corresponding compensator is strongly stabilizing, because conditions S1, S2 and S5 are satisfied.

- Let F be any strongly stabilizing feedback (for instance, $F = F_Q$) and choose $L = BK$, where K is statically stabilizing $\Sigma(A, B, C)$. Then the corresponding compensator is strongly stabilizing, as assumptions S1, S2 and S6 hold.

Very often, dynamic controllers are designed that render the closed-loop system input-output stable. A natural question is then whether the resulting closed-loop system is also strongly stable. In Theorem 6.3.2, we showed how to construct a strongly stabilizing compensator. The results that follow now give different conditions under which an input-output stabilizing compensator renders the closed-loop system strongly stable. These results are particularly applicable (but certainly not restricted) to the situation when a strongly stabilizable and detectable system is stabilized in the input-output sense by a finite-dimensional controller. In the remainder of this section, we do not assume that the direct feedthrough operator $D = 0$, in order to give an idea of how the formulas change in this case.

Let us consider the configuration of Figure 6.1, and let the systems Σ_G and Σ_K have the state-space realizations $\Sigma(A_G, B_G, C_G, D_G)$ and $\Sigma(A_K, B_K, C_K, D_K)$, respectively.

Theorem 6.3.3 *Consider the feedback interconnection as in Figure 6.1 of the systems* $\Sigma(A_G, B_G, C_G, D_G)$ *and* $\Sigma(A_K, B_K, C_K, D_K)$ *with transfer functions $G(s)$ and $K(s)$ and state spaces Z_G and Z_K, respectively. Assume that $K(s)$ is an admissible feedback transfer function for $G(s)$. Let the closed-loop system be described by the state-space realization* $\Sigma(A_{cl}, B_{cl}, C_{cl}, D_{cl})$ *on the product state space $Z_G \times Z_K$ and let G_{cl} denote the closed-loop transfer function.*

1. *If* $\Sigma(A_G, B_G, C_G, D_G)$ *and* $\Sigma(A_K, B_K, C_K, D_K)$ *are both strongly stabilizable, then* $\Sigma(A_{cl}, B_{cl}, C_{cl}, D_{cl})$ *is strongly stabilizable, as well.*

2. *If* $\Sigma(A_G, B_G, C_G, D_G)$ *and* $\Sigma(A_K, B_K, C_K, D_K)$ *are both strongly detectable, then* $\Sigma(A_{cl}, B_{cl}, C_{cl}, D_{cl})$ *is strongly detectable, as well.*

3. *If* $\Sigma(A_G, B_G, C_G, D_G)$ *and* $\Sigma(A_K, B_K, C_K, D_K)$ *are both strongly stabilizable and strongly detectable and $G_{cl} \in \mathbf{H}_\infty$, then* $\Sigma(A_{cl}, B_{cl}, C_{cl}, D_{cl})$ *is a strongly stable system.*

Proof It can be verified by straightforward calculations that the closed-loop system $\Sigma(A_{cl}, B_{cl}, C_{cl}, D_{cl})$ is given by

$$
A_{cl} = \begin{bmatrix} A_G & 0 \\ 0 & A_K \end{bmatrix} + \begin{bmatrix} B_G & 0 \\ 0 & B_K \end{bmatrix} \begin{bmatrix} -D_G & I \\ I & -D_K \end{bmatrix}^{-1} \begin{bmatrix} C_G & 0 \\ 0 & C_K \end{bmatrix},
$$

$$
B_{cl} = \begin{bmatrix} B_G & 0 \\ 0 & B_K \end{bmatrix} \begin{bmatrix} I & -D_K \\ -D_G & I \end{bmatrix}^{-1},
$$

$$
C_{cl} = \begin{bmatrix} I & -D_G \\ -D_K & I \end{bmatrix}^{-1} \begin{bmatrix} C_G & 0 \\ 0 & C_K \end{bmatrix},
$$

$$
D_{cl} = \begin{bmatrix} 0 & -I \\ -I & 0 \end{bmatrix} + \begin{bmatrix} -D_K & I \\ I & -D_G \end{bmatrix}^{-1}.
$$

1. By the stabilizability assumption, there exist $F_G \in \mathcal{L}(Z_G, U)$ and $F_K \in \mathcal{L}(Z_K, Y)$ such that $\Sigma(A_G + B_G F_G, 0, \begin{bmatrix} F_G^* & C_G^* \end{bmatrix}^*)$ and $\Sigma(A_K + B_K F_K, 0, \begin{bmatrix} F_K^* & C_K^* \end{bmatrix}^*)$ are strongly stable systems. It can now easily be verified that with

$$
F_{cl} = \begin{bmatrix} I & -D_K \\ -D_G & I \end{bmatrix} \left(\begin{bmatrix} F_G & 0 \\ 0 & F_K \end{bmatrix} - \begin{bmatrix} -D_G & I \\ I & -D_K \end{bmatrix}^{-1} \begin{bmatrix} C_G & 0 \\ 0 & C_K \end{bmatrix} \right),
$$

we obtain

$$A_{cl} + B_{cl}F_{cl} = \begin{bmatrix} A_G + B_G F_G & 0 \\ 0 & A_K + B_K F_K \end{bmatrix}$$

and $\Sigma(A_{cl} + B_{cl}F_{cl}, 0, [\begin{array}{cc} F_{cl}^* & C_{cl}^* \end{array}]^*)$ is a strongly stable system. Hence, the closed-loop system $\Sigma(A_{cl}, B_{cl}, C_{cl}, D_{cl})$ is strongly stabilizable.

2. By assumption, there exist operators $L_G \in \mathcal{L}(Y, Z_G)$ and $L_K \in \mathcal{L}(U, Z_K)$ such that $\Sigma(A_G + L_G C_G, [\begin{array}{cc} L_G & B_G \end{array}], 0)$ and $\Sigma(A_K + L_K C_K, [\begin{array}{cc} L_K & B_K \end{array}], 0)$ are strongly stable systems. It can now easily be verified that with

$$L_{cl} = \left(\begin{bmatrix} L_G & 0 \\ 0 & L_K \end{bmatrix} - \begin{bmatrix} B_G & 0 \\ 0 & B_K \end{bmatrix} \begin{bmatrix} -D_G & I \\ I & -D_K \end{bmatrix}^{-1} \right) \begin{bmatrix} I & -D_G \\ -D_K & I \end{bmatrix},$$

we obtain

$$A_{cl} + L_{cl}C_{cl} = \begin{bmatrix} A_G + L_G C_G & 0 \\ 0 & A_K + L_K C_K \end{bmatrix}$$

and $\Sigma(A_{cl} + L_{cl}C_{cl}, [\begin{array}{cc} L_{cl} & B_{cl} \end{array}], 0)$ is a strongly stable system. Hence, the closed-loop system $\Sigma(A_{cl}, B_{cl}, C_{cl}, D_{cl})$ is strongly detectable.

3. By parts 1 and 2 of this theorem, the resulting closed-loop system is strongly stabilizable and detectable. By assumption, it is input-output stable. Hence, we can apply Theorem 3.2.2 to prove that the closed-loop system is strongly stable.   □

In the literature, there are many results on stabilizing infinite-dimensional systems in an input-output sense by rational controllers which are usually (strongly) stabilizable and detectable. Theorem 6.3.3 gives testable conditions for internal asymptotic stability of the resulting closed-loop system.

We also have a version of the above theorem for the statically stabilizable case.

Theorem 6.3.4 *Suppose that the systems Σ_G and Σ_K are connected in feedback as in Figure 6.1 and that $I - D_G D_K$ is boundedly invertible. If both Σ_G and Σ_K are statically stabilizable, then the closed-loop system $\Sigma(A_{cl}, B_{cl}, C_{cl}, D_{cl})$ is statically stabilizable as well. If, in addition, the closed-loop system is input-output stable, then it is a strongly stable system.*

Proof Analogous to the proof of the previous theorem, it is easy to construct an output feedback operator that statically stabilizes the closed-loop system from the output feedback operators that statically stabilize the two subsystems Σ_G and Σ_K. The assertion about the strong stability of the closed-loop system follows from part 3 of Theorem 6.3.3 in combination with the fact that static stabilizability implies strong stabilizability and strong detectability.   □

As an application of these results we study the stability of the feedback interconnection of a system in the $\Sigma(A, B, B^*)$ class with a finite-dimensional controller. In the literature on flexible structures, one does consider strong stabilization (in the sense of obtaining a strongly stable closed-loop semigroup) by finite-dimensional strictly positive real dynamic compensators. Examples of this can be found in a number of papers by Morgül [75, 76, 77]. In these papers, ad-hoc methods are used to prove strong stability of the semigroup. The following result shows that in this case, the resulting closed-loop system is in fact a strongly stable system.

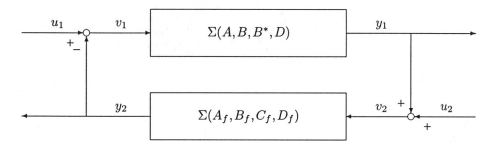

Figure 6.3: Negative feedback connection of $\Sigma(A, B, B^*, D)$ with a finite-dimensional strictly positive real compensator.

Theorem 6.3.5 *Consider the system $\Sigma(A, B, B^*, D)$ under assumptions A1–A5 of Section 2.2, connected in feedback with a finite-dimensional compensator $\Sigma(A_f, B_f, C_f, D_f)$ as in Figure 6.3. Assume, furthermore, that*

1. *the input space U is finite-dimensional,*

2. *$D + D^* \geq 0$,*

3. *$D_f + C_f(sI - A_f)^{-1}B_f$ is strictly positive real, and*

4. *$\Sigma(A_f, B_f, C_f, D_f)$ is stabilizable and detectable.*

Under these conditions, the closed-loop system is a strongly stable system.

Proof It is a well-known result that the negative feedback interconnection of a positive real system with a strictly positive real system is input-output stable (see, for instance, Desoer and Vidyasagar [39]). Now, $\Sigma(A, B, B^*, D)$ is statically stabilizable and $\Sigma(A_f, B_f, C_f, D_f)$ is stabilizable and detectable. Hence, part 3 of Theorem 6.3.3 applies to show that the closed-loop system is a strongly stable system. □

This result thus states that the interconnection of a system in the $\Sigma(A, B, B^*)$ class with a strictly positive real finite-dimensional stabilizable and detectable system leads to a strongly stable system. This is a very nice extension of the well-known fact that a system in the $\Sigma(A, B, B^*)$ class can be made strongly stable by a static negative definite output feedback.

Related results on input-output stabilization by dynamic compensators for a class of infinite-dimensional models for flexible structures can be found in Balakrishnan [7, 8, 9].

Chapter 7

Robust Stabilization

In this chapter, we consider the problem of robust stabilization of a strongly stabilizable linear system with respect to coprime factor perturbations. The configuration for stabilization that we consider is the one from Chapter 6: the plant and a dynamic compensator are connected in feedback as in Figure 6.1 and we aim at input-output stability of the closed-loop system. The controller, however, is not expected to stabilize just the plant: it should stabilize all plants which are close to the plant in the gap metric. That is, if $G = \tilde{M}^{-1}\tilde{N}$ is a left-coprime factorization, it should also stabilize all plants

$$G_\Delta = (\tilde{M} + \Delta_M)^{-1}(\tilde{N} + \Delta_N)$$

for any $\begin{bmatrix} \Delta_N & \Delta_M \end{bmatrix} \in \mathbf{H}_\infty$ having $\mathbf{H}_\infty$-norm less than some ε. Such a controller is called a robustly stabilizing controller with robustness margin ε. The problem is to find the maximal ε for which there exists such a controller and to find *all* controllers which stabilize G with a robustness margin at least equal to ε. A precise formulation of this problem will be given in Section 7.2

In both the finite-dimensional theory and the theory of exponentially stabilizable infinite-dimensional systems, the problem of robust stabilization with respect to coprime factor perturbations is reduced to a suboptimal Nehari problem and the parameterization of all robustly stabilizing controllers is derived from the parameterization of all solutions to the Nehari problem. We use the same approach and so we devote the next section to a treatment of the Nehari problem associated with nonexponentially stabilizable systems.

7.1 The Nehari Problem

The Nehari problem has received a lot of attention, both in the mathematical and the engineering literature (see Adamjan, Arov and Krein [1], Ball and Helton [10], Treil [102]). It has entered the control literature due to its connections with the parameterization of $\mathbf{H}_\infty$ controllers (see, for instance, Green et al. [52]).

For an operator-valued function $G \in \mathbf{L}_\infty(\mathcal{L}(U,Y))$, the *Nehari problem* is to find the minimum distance of G from the space of antistable operator-valued functions, i.e.,

$$\inf_{K(-s)\in\mathbf{H}_\infty(\mathcal{L}(U,Y))} \|G + K\|_\infty, \tag{7.1}$$

where U and Y are separable Hilbert spaces. It is well known that the infimum equals the norm of the Hankel operator with symbol G, $\|H_G\|$, which will be defined later in this section. (See Power [87], Partington [84] or Treil [102] for details on Hankel operators and the solution to the Nehari problem.)

The *suboptimal Nehari extension problem* is to find all $K(-s) \in \mathbf{H}_\infty(\mathcal{L}(U,Y))$ that satisfy

$$\|G + K\|_\infty \leq \sigma \tag{7.2}$$

for a given $\sigma > \|H_G\|$. This problem has been solved in the literature for any matrix-valued function $G \in \mathbf{L}_\infty$ (see Adamjan, Arov and Krein [1], Ball and Helton [10], Treil [102]). The solution in the frequency domain for $\|H_G\| < 1$ is given in terms of a linear-fractional parameter $\Theta = [\ \Theta_{11}\ \ \Theta_{12}\ \ \Theta_{21}\ \ \Theta_{22}\ ]$ with entries possibly only in $\mathbf{H}_2$. Moreover, this parameter arises from the J-spectral factorization $W = X^\sim J X$ for

$$W(s) = \begin{bmatrix} I & G(s) \\ 0 & I \end{bmatrix}^\sim \begin{bmatrix} I & 0 \\ 0 & -\sigma^2 I \end{bmatrix} \begin{bmatrix} I & G(s) \\ 0 & I \end{bmatrix}, \quad J = \begin{bmatrix} I & 0 \\ 0 & -I \end{bmatrix}. \tag{7.3}$$

For control applications one needs to give a state-space interpretation for the solutions. One supposes that $G(s)$ has a realization $G(s) = C(sI - A)^{-1}B$, where A, B, C are the generating operators of the state-space system, and one tries to find formulas for the solutions K in terms of these generating operators. These solutions then provide the parameterization of all stabilizing controllers for the $\mathbf{H}_\infty$ control problem. This has been carried out in Green et al. [52] for the case that A, B, C are matrices. This problem has also been studied in the literature, assuming that A is the generator of an exponentially stable C_0-semigroup and that B and C are linear operators. In Curtain and Zwart [36], U and Y are assumed to be finite-dimensional and B and C are bounded operators. Extensions to cases in which U and Y are infinite-dimensional or B and C are unbounded or both can be found in Glover, Curtain and Partington [50], Ran [88], Curtain and Ran [32], Foias and Tannenbaum [44], Curtain and Zwart [35] and Curtain and Ichikawa [25]. In all these papers, the assumption is made that A generates an exponentially stable C_0-semigroup and the Hankel operator is compact. However, we will study this problem for the case that $\Sigma(A, B, C)$ is not an exponentially stable system. Unlike in other chapters of this book, here we do not need to assume that $\Sigma(A, B, C)$ is strongly stable or strongly stabilizable. The specific class of systems we consider in this section are the systems $\Sigma(A, B, C)$ that satisfy the following hypotheses:

H1. A generates a C_0-semigroup on the separable Hilbert space Z and $B \in \mathcal{L}(U, Z)$, $C \in \mathcal{L}(Z, Y)$, where U and Y are finite-dimensional Hilbert spaces;

H2. $G(s) = C(sI - A)^{-1}B \in \mathbf{L}_\infty(\mathcal{L}(U, Y))$;

H3. $\Sigma(A, B, C)$ is input stable;

H4. $\Sigma(A, B, C)$ is output stable.

We remark that, instead of H2, we could have assumed that $G(s) \in \mathbf{H}_\infty(\mathcal{L}(U, Y))$ (i.e., $\Sigma(A, B, C)$ is input-output stable) because, by Lemma 4.2.4, the combination of input stability with $G \in \mathbf{L}_\infty(\mathcal{L}(U, Y))$ implies that $G \in \mathbf{H}_\infty(\mathcal{L}(U, Y))$.

Theorem 3.1 in Hansen and Weiss [53] (also see our Lemma 2.1.4) states that the system is input stable if and only if the Lyapunov equation

$$A\Pi_1 z + \Pi_1 A^* z = -BB^* z, \quad z \in \mathcal{D}(A^*), \tag{7.4}$$

has a self-adjoint, nonnegative solution (see also Grabowski [51]). In this case, one such solution is the *controllability Gramian* L_B of $\Sigma(A,B,C)$, which is defined as $L_B = \tilde{\Phi}\tilde{\Phi}^*$, where $\tilde{\Phi}$ is the extended input map (see (2.10)). By duality, the system is output stable if and only if the Lyapunov equation

$$A^*\Pi_2 z + \Pi_2 A z = -C^*Cz, \quad z \in \mathcal{D}(A), \tag{7.5}$$

has a self-adjoint, nonnegative solution. In this case, the *observability Gramian* L_C of $\Sigma(A,B,C)$, which is defined as $L_C = \Psi^*\Psi$, where Ψ is the extended output map, is one such self-adjoint, nonnegative solution of (7.5).

We remark that for the theory in this section, we do not require that the Lyapunov equations (7.4) and (7.5) have unique solutions; this will be the case if A^* (respectively, A) generates a strongly stable C_0-semigroup (see Lemma 2.1.4). However, we prefer not to make this assumption as it is unnecessary for our goal, which is to solve the Nehari problem (7.1) and its suboptimal extension in terms of the parameters A, B, C.

The main results of this section are the following two theorems.

Theorem 7.1.1 *Consider* $G(s) = C(sI - A)^{-1}B$ *under assumptions H1–H4. The following three statements are equivalent.*

1. *There exists a* $K(-s) \in \mathbf{H}_\infty(\mathcal{L}(U,Y))$ *such that*

$$\|G + K\|_\infty < \sigma. \tag{7.6}$$

2. $\sigma > \|H_G\| = r^{\frac{1}{2}}(L_B L_C)$, *where* r *denotes the spectral radius.*

3. *There exists an* $X(-s) \in \mathbf{H}_2(\mathcal{L}(Y \times U))$ *such that* $X^{-1}(-s) \in \mathbf{H}_2(\mathcal{L}(Y \times U))$ *and* $X_{11}^{-1}(-s) \in \mathbf{H}_\infty(\mathcal{L}(U))$, *which satisfies the factorization* $W = X^\sim JX$, *where* W *and* J *are given by (7.3).*

Theorem 7.1.2 *Let the assumptions H1–H4 be satisfied and let* $\sigma > r^{\frac{1}{2}}(L_B L_C)$. *All solutions* $K(-s) \in \mathbf{H}_\infty(\mathcal{L}(U,Y))$ *to the suboptimal Nehari extension problem (7.6) are given by*

$$K(-s) = R_1(-s)R_2(-s)^{-1}, \tag{7.7}$$

where

$$\begin{bmatrix} R_1(-s) \\ R_2(-s) \end{bmatrix} = X^{-1}(-s)\begin{bmatrix} Q(-s) \\ I_U \end{bmatrix} \tag{7.8}$$

and $Q(-s) \in \mathbf{H}_\infty(\mathcal{L}(U,Y))$ *satisfies* $\|Q\|_\infty \leq 1$.

To prove these theorems we have to develop some intermediate results. First we study the Hankel operator associated with our system. Hankel operators are at the core of the solution to the Nehari problem and we need to make the link between the frequency-domain Hankel operator and the time-domain Hankel operator. For a transfer function $G(s) = C(sI - A)^{-1}B \in \mathbf{L}_\infty(\mathcal{L}(U,Y))$, the *frequency-domain Hankel operator* $H_G : \mathbf{H}_2(U) \to \mathbf{H}_2(Y)$ is defined by

$$H_G f = \pi \Lambda_G f_-$$

for $f \in \mathbf{H}_2(U)$, where $f_-(s) := f(-s)$, Λ_G is the multiplication map induced by G and π is the orthogonal projection from $\mathbf{L}_2(-j\infty, j\infty; Y)$ onto $\mathbf{H}_2(Y)$. The *time-domain Hankel operator* $\Gamma : \mathbf{L}_2(0, \infty; U) \to \mathbf{L}_2(0, \infty; Y)$ is defined for $f \in \mathbf{L}_2(0, \infty; U)$ by

$$(\Gamma f)(t) = \int_0^\infty h(t+s)f(s)ds, \tag{7.9}$$

where we assume that the impulse response $h(\cdot) = CT(\cdot)B \in \mathbf{L}_2(0, \infty; \mathcal{L}(U, Y))$. This assumption is not sufficient to ensure that Γ is bounded, nor that Γ is isomorphic to H_G. We quote sufficient conditions from van Keulen [103].

Lemma 7.1.3 *Suppose that $h \in \mathbf{L}_2(0, \infty; \mathcal{L}(U, Y))$ and $\hat{h} \in \mathbf{L}_\infty(\mathcal{L}(U, Y))$, where $\hat{h}$ is the Laplace transform of h. Then, $\Gamma \in \mathcal{L}(\mathbf{L}_2(0, \infty; U), \mathbf{L}_2(0, \infty; Y))$ and*

$$\widehat{(\Gamma f)}(j\omega) = \left(H_{\hat{h}} \hat{f} \right)(j\omega)$$

for $f \in \mathbf{L}_2(0, \infty; U)$, where again ^ denotes the Laplace transform. Moreover, $\hat{h} \in \mathbf{H}_\infty(\mathcal{L}(U, Y))$.

The crucial condition in this result is that $\hat{h} \in \mathbf{L}_\infty$; in the absence of this condition it can happen that the time-domain Hankel operator Γ corresponds to a symbol different from $\hat{h}$ (see [103]).

We state the following results without proof. Some of them are easy to prove, the others are well known.

Lemma 7.1.4

1. *Let $f(s) = C(sI - A)^{-1}B$. f is holomorphic on $\mathbb{C}_0^+$ if and only if $f^\sim(s) := [f(-\bar{s})]^* = B^*(-sI - A^*)^{-1}C^*$ is holomorphic on $\mathbb{C}_0^-$. $f(s)$ is holomorphic on $\mathbb{C}_0^+$ if and only if $g(s) := f^\sim(-s) = B^*(sI - A^*)^{-1}C^*$ is holomorphic on $\mathbb{C}_0^+$. Moreover, $f \in \mathbf{H}_\infty$ if and only if $g \in \mathbf{H}_\infty$ and in this case $\|f\|_\infty = \|g\|_\infty$, and $f \in \mathbf{H}_2$ if and only if $g \in \mathbf{H}_2$ and in this case $\|f\|_2 = \|g\|_2$.*

2. *If $C(sI - A)^{-1}Bu \in \mathbf{H}_2(Y)$ for all $u \in U$ and U and Y are finite-dimensional, then $C(sI - A)^{-1}B \in \mathbf{H}_2(\mathcal{L}(U, Y))$.*

3. *If $f \in \mathbf{H}_\infty(\mathcal{L}(Z))$ and $g \in \mathbf{H}_2(\mathcal{L}(Z))$, then $fg \in \mathbf{H}_2(\mathcal{L}(Z))$ and $gf \in \mathbf{H}_2(\mathcal{L}(Z))$.*

Next, we prove a sequence of lemmas under the assumptions H1–H4. First, we apply Lemma 7.1.3 to our class of systems and make a link between the Hankel operators, the extended input and extended output maps and the controllability and observability Gramians, as introduced in (7.4) and (7.5).

Lemma 7.1.5 *Consider $\Sigma(A, B, C)$ under the assumptions H1–H4, let $h = CT(\cdot)B$ be its impulse response and let $G(s) = C(sI - A)^{-1}B$ be its transfer function.*

1. *$\Gamma = \Psi\tilde{\Phi}$, Γ is isomorphic to H_G, $G \in \mathbf{H}_\infty(\mathcal{L}(U, Y))$ and*

$$\|H_G\| = \|\Gamma\| = r^{\frac{1}{2}}(L_B L_C), \tag{7.10}$$

   where r denotes the spectral radius.

2. *$N_\sigma := (I - \frac{1}{\sigma^2} L_B L_C)^{-1} \in \mathcal{L}(Z)$ for $\sigma > \|\Gamma\|$.*

Proof 1.

$$\Psi(\tilde{\Phi}u)(t) = CT(t)\int_0^\infty T(s)Bu(s)ds$$

$$= \int_0^\infty CT(t+s)Bu(s)ds \quad \text{by the semigroup property}$$

$$= \int_0^\infty h(t+s)u(s)ds$$

$$= (\Gamma u)(t) \quad \text{by (7.9)}.$$

Thus, $\Gamma = \Psi\tilde{\Phi}$ is bounded, $h(\cdot)u \in \mathbf{L}_2(0,\infty;Y)$ for all $u \in U$ by H3 and H4 and $\hat{h}(s) = G(s) \in \mathbf{L}_\infty$ by H2 (see Chapter 4 in [36] for $\hat{h}(s) = G(s)$). So, Lemma 7.1.3 applies to show that Γ is isomorphic to H_G and $G \in \mathbf{H}_\infty(\mathcal{L}(U,Y))$. Finally,

$$\|H_G\|^2 = \|\Gamma\|^2 = \|\Gamma^*\Gamma\| = r(\Gamma^*\Gamma) \quad \text{since } \Gamma^*\Gamma \text{ is self-adjoint}$$
$$= r(\tilde{\Phi}^*\Psi^*\Psi\tilde{\Phi}) = r(\tilde{\Phi}\tilde{\Phi}^*\Psi^*\Psi) \quad \text{see Lemma A.4.15 in [36]}$$
$$= r(L_B L_C).$$

2. Note that $\sigma^2 > \|\Gamma\|^2 = r(L_B L_C)$ implies that the spectral radius of $\frac{1}{\sigma^2}L_B L_C$ is less than 1. Hence, $(I - \frac{1}{\sigma^2}L_B L_C)$ is boundedly invertible. $\square$

Next, we derive some results on two Riccati equations, which we need for the (crucial) Lemma 7.1.7.

Lemma 7.1.6 *Suppose that $\sigma > r^{\frac{1}{2}}(L_B L_C)$.*

1. *$W = N_\sigma L_B = L_B N_\sigma^* \in \mathcal{L}(Z)$ is a nonnegative solution of the following Riccati equation for $z \in \mathcal{D}(A^*)$:*

$$WA_W^* z + A_W Wz + \frac{1}{\sigma^2}WC^*CWz + N_\sigma BB^*N_\sigma^* z = 0, \qquad (7.11)$$

   where

$$A_W = A - \frac{1}{\sigma^2}N_\sigma L_B C^*C = A - \frac{1}{\sigma^2}WC^*C \qquad (7.12)$$

   generates the C_0-semigroup $T_W(t)$.

2. *$X = N_\sigma^* L_C = L_C N_\sigma$ is a nonnegative solution in $\mathcal{L}(Z)$ of the following Riccati equation for $z \in \mathcal{D}(A)$:*

$$A_X^* Xz + XA_X z + \frac{1}{\sigma^2}XBB^*Xz + N_\sigma^*C^*CN_\sigma z = 0, \qquad (7.13)$$

   where

$$A_X = A - \frac{1}{\sigma^2}BB^*L_C N_\sigma = A - \frac{1}{\sigma^2}BB^*X = N_\sigma^{-1}A_W N_\sigma \qquad (7.14)$$

   generates the C_0-semigroup $T_X(t)$.

3. *For all $z \in Z$, the functions $CW(sI - A_W^*)^{-1}z$, $CN_\sigma(sI - A_X)^{-1}z$ are in $\mathbf{H}_2(Y)$ and $B^*N_\sigma^*(sI - A_W^*)^{-1}z$, $B^*X(sI - A_X)^{-1}z$ are in $\mathbf{H}_2(U)$.*

Proof Parts 1 and 2. First, we need to show that $WD(A^*) \subset D(A)$ and $XD(A) \subset D(A^*)$. The proofs of Lemmas 4.1.24 and 8.3.2 in [36] apply here to show that $L_B D(A^*) \subset D(A)$, $L_C D(A) \subset D(A^*)$, $N_\sigma D(A) \subset D(A)$ and $N_\sigma^* D(A^*) \subset D(A^*)$. That W and X satisfy (7.11) and (7.13), respectively, is then a matter of algebra using the Lyapunov equations (7.4) and (7.5).

To show that $W \geq 0$, consider

$$
\begin{aligned}
\langle Wz, z \rangle &= \left\langle L_B \left(I - \frac{1}{\sigma^2} L_C L_B \right)^{-1} z, z \right\rangle \\
&= \left\langle L_B z_0, \left(I - \frac{1}{\sigma^2} L_C L_B \right) z_0 \right\rangle, \quad \text{where } z_0 = N_\sigma^{-*} z \\
&= \langle \tilde{\Phi}^* z_0, \tilde{\Phi}^* z_0 \rangle - \frac{1}{\sigma^2} \langle \Psi \tilde{\Phi} \tilde{\Phi}^* z_0, \Psi \tilde{\Phi} \tilde{\Phi}^* z_0 \rangle \quad \text{by (7.4), (7.5), Lemma 7.1.5} \\
&= \| \tilde{\Phi}^* z_0 \|^2 - \frac{1}{\sigma^2} \| \Gamma \tilde{\Phi}^* z_0 \|^2 \\
&\geq 0 \quad \text{since } \sigma > \|\Gamma\|.
\end{aligned}
$$

The proof that $X \geq 0$ is analogous.

3. We prove only the W-results, as those for X are analogous. Appealing to the Paley–Wiener theorem (see, e.g., Theorem A.6.21 in [36]), it is sufficient to derive the time-domain inequalities

$$
\int_0^\infty \| C W T_W^*(t) z_1 \|^2 dt \leq \sigma^2 \|W\| \|z_1\|^2, \tag{7.15}
$$

$$
\int_0^\infty \| B^* N_\sigma^* T_W^*(t) z_1 \|^2 dt \leq \|W\| \|z_1\|^2.
$$

These follow from (7.11) substituting $z = T_W^*(t) z_1$ for $z_1 \in D(A^*)$ and taking inner products with z to obtain

$$
\frac{d}{dt} \langle T_W^*(t) z_1, W T_W^*(t) z_1 \rangle = -\frac{1}{\sigma^2} \| C W T_W^*(t) z_1 \|^2 - \| B^* N_\sigma^* T_W^*(t) z_1 \|^2.
$$

Integrating from 0 to t yields

$$
\begin{aligned}
\langle z_1, W z_1 \rangle &= \langle T_W^*(t) z_1, W T_W^*(t) z_1 \rangle + \frac{1}{\sigma^2} \int_0^t \| C W T_W^*(s) z_1 \|^2 ds \\
&\quad + \int_0^t \| B^* N_\sigma^* T_W^*(s) z_1 \|^2 ds.
\end{aligned}
$$

Since $W \geq 0$ and $D(A^*)$ is dense in Z, this proves (7.15). $\qquad\square$

Most of the recent papers on the suboptimal Nehari extension problem exploit the relation with a certain J-spectral factorization. We will do the same and in the following lemma we give a condition for the existence of the factorization and derive a number of properties of the spectral factor.

Lemma 7.1.7 *Let $G(s) = C(sI - A)^{-1}B$, and assume that conditions H1–H4 are satisfied. Let $\sigma > r^{\frac{1}{2}}(L_B L_C)$ and define $X(s)$ by*

$$
X(s) = \begin{bmatrix} I_Y & 0 \\ 0 & \sigma I_U \end{bmatrix} + \frac{1}{\sigma^2} \begin{bmatrix} -C L_B \\ \sigma B^* \end{bmatrix} N_\sigma^*(sI + A^*)^{-1} \begin{bmatrix} C^* & L_C B \end{bmatrix}. \tag{7.16}
$$

X has the following properties.

1. X satisfies the factorization for $\omega \in \mathbb{R}$,

$$W(j\omega) = X^*(j\omega) \begin{bmatrix} I_Y & 0 \\ 0 & -I_U \end{bmatrix} X(j\omega), \qquad (7.17)$$

where

$$W(j\omega) = \begin{bmatrix} I_Y & G(j\omega) \\ 0 & I_U \end{bmatrix}^* \begin{bmatrix} I_Y & 0 \\ 0 & -\sigma^2 I_U \end{bmatrix} \begin{bmatrix} I_Y & G(j\omega) \\ 0 & I_U \end{bmatrix}.$$

2. $X(s)$ is invertible and its inverse $V(s)$ is given by

$$V(s) = \begin{bmatrix} I_Y & 0 \\ 0 & \frac{1}{\sigma} I_U \end{bmatrix} - \frac{1}{\sigma^2} \begin{bmatrix} -CL_B \\ B^* \end{bmatrix} (sI + A^*)^{-1} N_\sigma^* \begin{bmatrix} C^* & \frac{1}{\sigma} L_C B \end{bmatrix}. \quad (7.18)$$

3. $X(-s) - \begin{bmatrix} I_Y & 0 \\ 0 & \sigma I_U \end{bmatrix} \in \mathbf{H}_2(\mathcal{L}(Y \times U))$.

4. $V(-s) - \begin{bmatrix} I_Y & 0 \\ 0 & \frac{1}{\sigma} I_U \end{bmatrix} \in \mathbf{H}_2(\mathcal{L}(Y \times U))$.

5. $X_{11}(-s)$ is invertible and its inverse is given by

$$X_{11}^{-1}(-s) - I = -\frac{1}{\sigma} CW(sI - A_W^*)^{-1} C^* \in \mathbf{H}_2 \cap \mathbf{H}_\infty(\mathcal{L}(Y)).$$

6. $V_{22}(-s)$ is invertible and its inverse is given by

$$V_{22}^{-1}(-s) - \sigma I = -\frac{1}{\sigma} B^*(sI - A_X^*)^{-1} XB \in \mathbf{H}_2 \cap \mathbf{H}_\infty(\mathcal{L}(U)).$$

7. $X_{21}(-s)X_{11}^{-1}(-s) = -V_{22}^{-1}(-s)V_{21}(-s) \in \mathbf{H}_2 \cap \mathbf{H}_\infty(\mathcal{L}(Y,U))$ and is given by

$$X_{21}(-s)X_{11}^{-1}(-s) = -\frac{1}{\sigma} B^* N_\sigma^*(sI - A_W^*)^{-1} C^*$$

$$= -\frac{1}{\sigma} B^*(sI - A_X^*)^{-1} N_\sigma^* C^*.$$

8. $X_{11}^{-1}(-s)X_{12}(-s) = -V_{12}(-s)V_{22}^{-1}(-s) \in \mathbf{H}_2 \cap \mathbf{H}_\infty(\mathcal{L}(U,Y))$ and is given by

$$X_{11}^{-1}(-s)X_{12}(-s) = \frac{1}{\sigma^2} CW(sI - A_W^*)^{-1} L_C B$$

$$= \frac{1}{\sigma^2} CL_B(sI - A_X^*)^{-1} XB.$$

Proof The proofs of all the explicit formulas follow by algebraic manipulations using the Lyapunov identities (7.4), (7.5) and the Riccati equations (7.11), (7.13), noting that $L_B \mathcal{D}(A^*) \subset \mathcal{D}(A)$, etc. For details see [36, Corollary 7.3.7 and Lemma 8.3.3]. As an example, we calculate

$$V_{22}^{-1}(-s) = \sigma \left(I_U + \frac{1}{\sigma^2} B^*(sI - A^*)^{-1} N_\sigma^* L_C B \right)^{-1}$$

$$= \sigma \left(I_U - \frac{1}{\sigma^2} B^* \left(sI - A^* + \frac{1}{\sigma^2} XBB^* \right)^{-1} N_\sigma^* L_C B \right)$$

$$= \sigma I_U - \frac{1}{\sigma} B^*(sI - A_X^*)^{-1} XB,$$

where we use Corollary 7.3.7 from [36] for the second equality. Then,

$$
\begin{aligned}
V_{22}^{-1}(-s)V_{21}(-s) &= \sigma\left(I_U - \frac{1}{\sigma^2}B^*(sI - A_X^*)^{-1}XB\right)\frac{1}{\sigma^2}B^*(sI - A^*)^{-1}N_\sigma^*C^* \\
&= \frac{1}{\sigma}B^*(sI - A_X^*)^{-1}\left\{sI - A_X^* - \frac{1}{\sigma^2}XBB^*\right\}(sI - A^*)^{-1}N_\sigma^*C^* \\
&= \frac{1}{\sigma}B^*(sI - A_X^*)^{-1}N_\sigma^*C^* \\
&= \frac{1}{\sigma}B^*N_\sigma^*(sI - A_W^*)^{-1}C^* \quad \text{using (7.14);}
\end{aligned}
$$

$$
\begin{aligned}
V_{12}(-s)V_{22}^{-1}(-s) &= -\sigma^{-3}CL_B(sI - A^*)^{-1}N_\sigma^*L_CB\sigma\left(I_U - \frac{1}{\sigma^2}B^*(sI - A_X^*)^{-1}XB\right) \\
&= -\frac{1}{\sigma^2}CL_B(sI - A^*)^{-1}\left\{sI - A_X^* - \frac{1}{\sigma^2}XBB^*\right\}(sI - A_X^*)^{-1}XB \\
&= -\frac{1}{\sigma^2}CL_B(sI - A_X^*)^{-1}XB \\
&= -\frac{1}{\sigma^2}CW(sI - A_W^*)^{-1}L_CB \quad \text{using (7.14).}
\end{aligned}
$$

The formulas for X_{11}^{-1}, $X_{21}X_{11}^{-1}$ and $X_{11}^{-1}X_{12}$ can be calculated in a similar fashion, yielding the identities asserted in parts 7 and 8.

The tricky part is to show that the transfer functions are in the right spaces. From H4 we obtain that $X_{11}(-s) - I \in \mathbf{H}_2(\mathcal{L}(Y))$ and $X_{21}(-s) \in \mathbf{H}_2(\mathcal{L}(Y,U))$. It follows from H3 that $V_{21}(-s) \in \mathbf{H}_2(\mathcal{L}(Y,U))$ and $V_{22}(-s) - \frac{1}{\sigma}I \in \mathbf{H}_2(\mathcal{L}(U))$. For this, we use part 2 of Lemma 7.1.6. Using part 3 of Lemma 7.1.6, we deduce that $V_{22}^{-1}(-s) - \sigma I \in \mathbf{H}_2(\mathcal{L}(U))$, $X_{11}^{-1}(-s) - I \in \mathbf{H}_2(\mathcal{L}(Y))$, and the functions in parts 7 and 8 are in $\mathbf{H}_2(\mathcal{L}(Y,U))$ and $\mathbf{H}_2(\mathcal{L}(U,Y))$, respectively.

To prove that the functions in parts 5, 6, 7 and 8 are also in $\mathbf{H}_\infty$, we first deduce some inequalities from the factorization (7.17) and its inverse relation

$$
\begin{bmatrix} I_Y & -G \\ 0 & I_U \end{bmatrix}\begin{bmatrix} I_Y & 0 \\ 0 & -\sigma^2 I_U \end{bmatrix}\begin{bmatrix} I_Y & -G \\ 0 & I_U \end{bmatrix}^* = V\begin{bmatrix} I_Y & 0 \\ 0 & -I_U \end{bmatrix}V^*. \tag{7.19}
$$

The $(1,1)$-block of (7.17) yields

$$
X_{11}(j\omega)^*X_{11}(j\omega) - X_{21}(j\omega)^*X_{21}(j\omega) = I,
$$

whence, $X_{11}^{-1} \in \mathbf{L}_\infty$, $X_{21}X_{11}^{-1} \in \mathbf{L}_\infty$ and

$$
\|X_{11}^{-1}\|_\infty \leq 1, \quad \|X_{21}X_{11}^{-1}\|_\infty < 1. \tag{7.20}
$$

The $(2,2)$-block of (7.19) yields

$$
V_{21}(j\omega)V_{21}(j\omega)^* - V_{22}(j\omega)V_{22}(j\omega)^* = -\frac{1}{\sigma^2}I,
$$

whence, $V_{22}^{-1} \in \mathbf{L}_\infty$, $V_{22}^{-1}V_{21} \in \mathbf{L}_\infty$ and

$$\|V_{22}^{-1}\|_\infty \le \sigma, \quad \|V_{22}^{-1}V_{21}\|_\infty < 1. \tag{7.21}$$

Now, Lemma 4.2.4 applies to show that $X_{11}^{-1}(-s) \in \mathbf{H}_\infty(\mathcal{L}(Y))$, $V_{22}^{-1}(-s) \in \mathbf{H}_\infty(\mathcal{L}(U))$ and the functions in part 7 are in $\mathbf{H}_\infty(\mathcal{L}(Y,U))$. To show that $V_{12}(-s)V_{22}^{-1}(-s) = X_{11}^{-1}(-s)X_{12}(-s) \in \mathbf{H}_\infty(\mathcal{L}(U,Y))$, we introduce $K(-s) = V_{12}(-s)V_{22}^{-1}(-s)$. Then,

$$\begin{bmatrix} G+K \\ I \end{bmatrix} = \begin{bmatrix} I & G \\ 0 & I \end{bmatrix} V \begin{bmatrix} 0 \\ V_{22}^{-1} \end{bmatrix},$$

and for $s = j\omega$, $\omega \in \mathbb{R}$, we have

$$(G+K)^*(G+K) - \sigma^2 I \tag{7.22}$$

$$= \begin{bmatrix} G+K \\ I \end{bmatrix}^* \begin{bmatrix} I & 0 \\ 0 & -\sigma^2 I \end{bmatrix} \begin{bmatrix} G+K \\ I \end{bmatrix}$$

$$= \begin{bmatrix} 0 \\ V_{22}^{-1} \end{bmatrix}^* V^* \begin{bmatrix} I & 0 \\ G^* & I \end{bmatrix} \begin{bmatrix} I & 0 \\ 0 & -\sigma^2 I \end{bmatrix} \begin{bmatrix} I & G \\ 0 & I \end{bmatrix} V \begin{bmatrix} 0 \\ V_{22}^{-1} \end{bmatrix}$$

$$= \begin{bmatrix} 0 \\ V_{22}^{-1} \end{bmatrix}^* \begin{bmatrix} I & 0 \\ 0 & -I \end{bmatrix} \begin{bmatrix} 0 \\ V_{22}^{-1} \end{bmatrix} < 0,$$

where we have used the spectral factorization. Thus, $G+K \in \mathbf{L}_\infty$ and $\|G+K\|_\infty < \sigma$. Consequently, $K \in \mathbf{L}_\infty$ and $\|K\|_\infty < \|G\|_\infty + \sigma < \infty$. Previously, we proved that $K(-s) \in \mathbf{H}_2(\mathcal{L}(U,Y))$. So, Lemma 4.2.4 applies to show that $K(-s) \in \mathbf{H}_\infty(\mathcal{L}(U,Y))$.

Finally, we must show that $X_{12}(-s)$, $X_{22}(-s)$, $V_{11}(-s)$ and $V_{12}(-s) \in \mathbf{H}_2$. From the identity $XV = VX = I$, we obtain

$$V_{11}(-s) = X_{11}^{-1}(-s) - \{X_{11}^{-1}(-s)X_{12}(-s)\} V_{12}(-s),$$
$$X_{22}(-s) = V_{22}^{-1}(-s) - X_{21}(-s)\{V_{12}(-s)V_{22}^{-1}(-s)\}.$$

Furthermore,

$$X_{12}(-s) = X_{11}\{X_{11}^{-1}X_{12}(-s)\},$$
$$V_{12}(-s) = [V_{12}V_{22}^{-1}] V_{22}(-s).$$

The results now follow, using part 3 of Lemma 7.1.4 and the fact that the terms between curly braces are in $\mathbf{H}_2 \cap \mathbf{H}_\infty$ and the others in $\mathbf{H}_2$. □

Although the above proof shows many similarities to proofs in Curtain and Zwart [35, 36], there is a significant difference. The exponential stability assumption made in [35, 36], means that it is fairly easy to show that $X(-s)$ and $V(-s)$ are in $\mathbf{H}_\infty(\mathcal{L}(Y \times U))$. In our case, however, this assumption no longer holds and we obtain only $V(-s)$, $X(-s) \in \mathbf{H}_2(\mathcal{L}(Y \times U))$ and that some components are in $\mathbf{H}_\infty$. Fortunately, these are the crucial ones we need. Because X and V are not in $\mathbf{H}_\infty$, the factorization in (7.17) is *not* a J-spectral factorization, according to the usual definitions.

One would expect that the above results are also valid even if U and Y are infinite-dimensional. However, in this case the situation becomes very complicated. We were able to prove that $V(-s)(u,y) \in \mathbf{H}_2(Y \times U)$ for all $u \in U$ and $y \in Y$, but we could not prove a similar claim for $X(-s)$. The main reason for this is that part 2 of Lemma 7.1.4 is not valid in this case, and that the second claim in part 3 of Lemma 7.1.4 does not extend to functions $g(s)$ that satisfy only $g(s)z \in \mathbf{H}_2(Z)$ for all $z \in Z$.

Now we are able to prove our main results.

Proof of Theorem 7.1.1 1 $\Rightarrow$ 2. It is well known (see Treil [102]) that

$$\inf_{K(-s)\in\mathbf{H}_\infty(\mathcal{L}(U,Y))} \|G+K\|_\infty = \|H_G\|.$$

In Lemma 7.1.5, it was shown that $\|H_G\| = r^{\frac{1}{2}}(L_B L_C)$. Combining the two, we obtain statement 2.

2 $\Rightarrow$ 3. We have shown this in Lemma 7.1.7.

3 $\Rightarrow$ 1. We take $K(-s) = V_{12}(-s)V_{22}^{-1}(-s)$. In the proof of Lemma 7.1.7, this K was shown to satisfy $K(-s) \in \mathbf{H}_\infty(\mathcal{L}(U,Y))$ and $\|G+K\|_\infty < \sigma$.

Proof of Theorem 7.1.2 Suppose that $K(-s)$ is given by (7.7) and (7.8) for some $Q \in \mathbf{H}_\infty(\mathcal{L}(U,Y))$, $\|Q\|_\infty \leq 1$. In the proof of Lemma 7.1.7, we showed that $V_{22}^{-1}V_{21}$ satisfies $\|V_{22}^{-1}V_{21}\|_\infty < 1$. Hence, $\|V_{22}^{-1}V_{21}Q\|_\infty < 1$ and $(I+V_{22}^{-1}V_{21}Q)^{-1} \in \mathbf{H}_\infty(\mathcal{L}(U))$, because $\mathbf{H}_\infty$ is a Banach algebra. Consequently,

$$R_2^{-1} = (I+V_{22}^{-1}V_{21}Q)^{-1}V_{22}^{-1} \in \mathbf{H}_\infty(\mathcal{L}(U)).$$

In addition, by part 4 of Lemma 7.1.7 and part 3 of Lemma 7.1.4,

$$\begin{bmatrix} R_1(-s) \\ R_2(-s) \end{bmatrix} - \begin{bmatrix} Q(-s) \\ \frac{1}{\sigma}I \end{bmatrix} \in \mathbf{H}_2(\mathcal{L}(U,U\times Y)).$$

So,

$$\begin{aligned} K(-s) &= R_1(-s)R_2^{-1}(-s) \\ &= [R_1(-s)-Q(-s)]R_2^{-1}(-s) + Q(-s)R_2^{-1}(-s) \end{aligned}$$

is the sum of an $\mathbf{H}_2$- and an $\mathbf{H}_\infty$-function, because $(R_1(-s)-Q(-s)) \in \mathbf{H}_2(\mathcal{L}(U))$ and both $R_2^{-1}(-s)$ and $Q(-s) \in \mathbf{H}_\infty(\mathcal{L}(U))$. We can show analogously to the derivation in (7.22) that on the imaginary axis,

$$(G+K)^*(G+K) - \sigma^2 I = (R_2^{-1})^*(Q^*Q-I)R_2^{-1} \leq 0,$$

and so $G+K \in \mathbf{L}_\infty$ and, consequently, $K \in \mathbf{L}_\infty$. Moreover, this inequality shows that all K given by (7.7) and (7.8) satisfy $\|G+K\| < \sigma$. Lemma 4.2.4 applies to show that $K(-s) \in \mathbf{H}_\infty(\mathcal{L}(U,Y))$.

For the converse, suppose that $K(-s) \in \mathbf{H}_\infty(\mathcal{L}(U,Y))$ satisfies (7.2). We seek a representation $K(-s) = R_1(-s)R_2(-s)^{-1}$. Define

$$\begin{bmatrix} U_1(-s) \\ U_2(-s) \end{bmatrix} := X(-s)\begin{bmatrix} K(-s) \\ I_U \end{bmatrix}. \tag{7.23}$$

From part 3 of Lemma 7.1.4 and part 3 of Lemma 7.1.7, we deduce that both $U_1(-s)$ and $U_2(-s)$ are sums of $\mathbf{H}_2$- and $\mathbf{H}_\infty$-functions. As in the proof of Theorem 8.3.7 of [36], we can deduce the following results:

$$\|U_1(j\omega)u\|_Y \leq \|U_2(j\omega)u\|_U \quad \text{for } u \in U, \omega \in \mathbb{R}, \tag{7.24}$$

$U_2(j\omega)$ is invertible over $\mathcal{L}(U)$ for all $\omega \in \mathbb{R}$ and

$$U(j\omega) := U_1(j\omega)U_2^{-1}(j\omega) \in \mathbf{L}_\infty(\mathcal{L}(U,Y)) \tag{7.25}$$

satisfies

$$\|U\|_\infty \leq 1. \tag{7.26}$$

We now identify $R_2(-s) = U_2^{-1}(-s)$, $R_1(-s) = K(-s)R_2(-s)$, which give an appropriate representation,

$$K(-s) = R_1(-s)R_2^{-1}(-s), \tag{7.27}$$

$$\begin{bmatrix} R_1(-s) \\ R_2(-s) \end{bmatrix} = X^{-1}(-s)\begin{bmatrix} U(-s) \\ I_U \end{bmatrix}. \tag{7.28}$$

It remains to prove that $U(-s) \in \mathbf{H}_\infty(\mathcal{L}(U,Y))$. Define

$$F(s) = -V_{22}^{-1}(-s)V_{21}(-s)U(-s).$$

F is in $\mathbf{L}_\infty(\mathcal{L}(U))$ by (7.21) and (7.26) and $\|F\|_\infty < 1$. Because $\mathbf{L}_\infty(\mathcal{L}(U))$ is a Banach algebra, we can conclude that $I - F$ is invertible over $\mathbf{L}_\infty$ and using the representation (7.23), it is easily checked that

$$(I - F(s))^{-1} = U_2(-s)V_{22}(-s) \in \mathbf{L}_\infty(\mathcal{L}(U)).$$

Thus,

$$U_2(-s) = (I - F(s))^{-1}V_{22}^{-1}(-s) \in \mathbf{L}_\infty(\mathcal{L}(U)),$$

and (7.24) shows that $U_1(-s) \in \mathbf{L}_\infty(\mathcal{L}(U,Y))$, as well. We have already proved that $U_1(-s)$ and $U_2(-s)$ are the sums of $\mathbf{H}_2$- and $\mathbf{H}_\infty$-functions and so by Lemma 4.2.4, we conclude that they are both in $\mathbf{H}_\infty$. Thus,

$$\begin{aligned} (I - F(s))^{-1} &= U_2(-s)V_{22}(-s) \\ &= U_2(-s)\{V_{22}(-s) - \sigma I\} + \sigma U_2(-s) \end{aligned}$$

is the sum of a function in $\mathbf{H}_2$ and a function in $\mathbf{H}_\infty$. This, together with $(I - F(s))^{-1} \in \mathbf{L}_\infty$, shows that it is also in $\mathbf{H}_\infty$, again using Lemma 4.2.4. By Lemma 8.3.5 of Curtain and Zwart [36], we obtain that $F(s) \in \mathbf{H}_\infty(\mathcal{L}(U))$. As $V_{22}(-s) - \sigma^{-1}I \in \mathbf{H}_2(\mathcal{L}(U))$,

$$U_2^{-1}(-s) = V_{22}(-s)(I - F(s))$$

is the sum of a function in $\mathbf{H}_2$ and a function in $\mathbf{H}_\infty$, and as $U_1 \in \mathbf{H}_\infty$,

$$U(-s) = U_1(-s)U_2^{-1}(-s)$$

is the sum of a function in $\mathbf{H}_2$ and a function in $\mathbf{H}_\infty$, as well. However, we have already proved that $U(-s) \in \mathbf{L}_\infty(\mathcal{L}(U,Y))$ and so we can apply Lemma 4.2.4 to obtain that $U(-s) \in \mathbf{H}_\infty(\mathcal{L}(U,Y))$. □

7.2   Parameterization of Robustly Stabilizing Controllers

In this section we formulate the problem of robust stabilization and we present a parameterization of controllers that solve the problem. We consider the robust stabilization problem for those systems for which we obtained explicit formulas for a normalized doubly coprime factorization in Chapter 6. More explicitly, we consider systems $\Sigma(A, B, C, D)$ with finite-dimensional input and output spaces that satisfy assumptions (i)–(vi) of Theorem 6.2.4:

(i) A has no essential spectrum on the imaginary axis;

(ii) there exists an $F \in \mathcal{L}(Z, \mathbb{C}^m)$ such that $\begin{bmatrix} F^* & C^* \end{bmatrix}^* (sI - A_F)^{-1} z \in \mathbf{H}_2(U \times Y)$ for all $z \in Z$;

(iii) $\Sigma(A, B, C, D)$ is strongly detectable;

(iv) A^* has no essential spectrum on the imaginary axis;

(v) there exists an $L \in \mathcal{L}(\mathbb{C}^k, Z)$ such that $\begin{bmatrix} B & L \end{bmatrix}^* (sI - A_L^*)^{-1} z \in \mathbf{H}_2(U \times Y)$ for all $z \in Z$;

(vi) $\Sigma(A^*, C^*, B^*, D^*)$ is strongly detectable.

A class of systems satisfying these assumptions is that of strongly stabilizable and detectable systems $\Sigma(A, B, C, D)$, with finite-dimensional inputs and outputs and where A has compact resolvent. An important special case is that of systems in the $\Sigma(A, B, B^*)$ class with finite-dimensional input space. The extra conditions in Theorem 6.2.8 that allowed us to relieve the assumption of a finite-dimensional input space are of no help in this chapter: the restriction on the dimension of U is necessary because we will apply the results on the Nehari problem of the previous section, and those are only valid for finite-dimensional input and output spaces.

The systems in our class have transfer matrices that have entries in the quotient field $\mathbf{H}_\infty[\mathbf{H}_\infty]^{-1}$. We will denote the class of matrices of any size with components in the field $\mathcal{F}$ by $\mathcal{MF}$.

For a plant with a transfer matrix in $\mathcal{M}\mathbf{H}_\infty[\mathbf{H}_\infty]^{-1}$, possessing a normalized doubly coprime factorization, we consider the following class of perturbations.

Definition 7.2.1 *Let G have a left-coprime factorization $(\tilde{M}, \tilde{N})$. ε-admissible left-coprime factor perturbations of G are plants with transfer matrices given by*

$$G_\Delta = (\tilde{M} + \Delta_M)^{-1}(\tilde{N} + \Delta_N),$$

where $\Delta = \begin{bmatrix} \Delta_N & -\Delta_M \end{bmatrix} \in \mathcal{M}\mathbf{H}_\infty$ and $\|\Delta\|_\infty < \varepsilon$.

We consider the *robust stabilization problem* for G: find all $K \in \mathcal{M}\mathbf{H}_\infty[\mathbf{H}_\infty]^{-1}$ that stabilize all ε-admissible perturbations of G for a given $\varepsilon > 0$. In addition, we are interested in finding the maximal robustness margin ε_{max}, i.e., the largest ε for which the above problem has a solution. In this problem formulation, as well as in the remainder of this section, stabilization will mean input-output stabilization.

The problem of robust stabilization with respect to normalized coprime factor perturbations is now a classic $\mathbf{H}_\infty$ control problem that admits an elegant solution via the solution of a Nehari problem for a normalized left-coprime factorization of the system transfer function. The most general solution was obtained by Georgiou and Smith in [47] for plants possessing a normalized left-coprime factorization. It is known that not all transfer functions possess coprime factorizations, but that the existence of an input-output stabilizing transfer function is a necessary and sufficient condition for this (see Smith [95]). The disadvantage of the latter solution is that, although very general, it does not yield explicit formulas for the robustly stabilizing controllers. These depend on explicit formulas for a normalized doubly-coprime factorization of the transfer function and these are not available for irrational transfer functions, in general. Special classes for which these are available can be found in Curtain [23], Curtain, Weiss and Weiss [34] and Curtain and Zwart [37]. Furthermore, we obtained them for a class of strongly stabilizable and detectable systems in Chapter 6.

In McFarlane and Glover [72], known formulas for the normalized coprime factorizations were exploited to obtain explicit formulas for the robustly stabilizing controllers in terms of the control and filter Riccati equations of the original system. This elegant finite-dimensional result was extended to infinite-dimensional systems with bounded inputs and outputs in Section 9.4 of Curtain and Zwart [36] and to systems with unbounded inputs and outputs in Curtain [23] and Curtain and Pritchard [31]. All of these results assume that the state-space realization is exponentially stabilizable and detectable. A different frequency-domain approach was used by Dym, Georgiou and Smith [43] to solve the problem for a special class of delay systems (see [43]).

Our approach is to adapt the theory of Chapter 9 of Curtain and Zwart [36] to our situation, where we replace exponential stability (stabilizability) with strong stability (stabilizability) and the Callier–Desoer algebra of transfer functions $\hat{\mathcal{A}}_-$ with the algebra $\mathbf{H}_\infty$.

First, we translate the robust stabilization problem into an $\mathbf{H}_\infty$-optimization problem, as in Section 4.1 of McFarlane and Glover [72] and Corollary 9.2.9 of Curtain and Zwart [36]. In a later stage, this optimization problem will be related to a Nehari extension problem.

Theorem 7.2.2 *Suppose that $G \in \mathcal{M}\mathbf{H}_\infty[\mathbf{H}_\infty]^{-1}$ has a normalized doubly coprime factorization over $\mathbf{H}_\infty$. Let $0 < \varepsilon < 1$ and $K \in \mathcal{M}\mathbf{H}_\infty[\mathbf{H}_\infty]^{-1}$. If K stabilizes G and*

$$\left\| \begin{bmatrix} K \\ I \end{bmatrix} (I - GK)^{-1} \tilde{M}^{-1} \right\|_\infty \leq \frac{1}{\varepsilon}, \tag{7.29}$$

then it stabilizes all ε-admissible left-coprime factor perturbations of G.

Proof Let us introduce the shorthand notation

$$F(s) = \begin{bmatrix} K(s) \\ I \end{bmatrix} (I - G(s)K(s))^{-1} \tilde{M}^{-1}(s).$$

Using $G = \tilde{M}^{-1}\tilde{N}$ and $K = UV^{-1}$, we have

$$F = \begin{bmatrix} U \\ V \end{bmatrix} (\tilde{M}V - \tilde{N}U)^{-1}.$$

By Lemma 6.1.4, $F \in \mathbf{H}_\infty(\mathcal{L}(Y, U \times Y))$ and by assumption $\|F\|_\infty \leq \varepsilon^{-1}$. Now,

$$\begin{aligned}
(I - G_\Delta K) &= I - (\tilde{M} + \Delta_M)^{-1}(\tilde{N} + \Delta_N)UV^{-1} \\
&= (\tilde{M} + \Delta_M)^{-1} \left((\tilde{M} + \Delta_M)V - (\tilde{N} + \Delta_N)U \right) V^{-1} \\
&= (\tilde{M} + \Delta_M)^{-1} \left(\tilde{M}V - \tilde{N}U - [\, \Delta_N \quad -\Delta_M \,] \begin{bmatrix} U \\ V \end{bmatrix} \right) V^{-1} \\
&= (\tilde{M} + \Delta_M)^{-1} \left(I - \Delta \begin{bmatrix} U \\ V \end{bmatrix} (\tilde{M}V - \tilde{N}U)^{-1} \right) (\tilde{M}V - \tilde{N}U) V^{-1} \\
&= (\tilde{M} + \Delta_M)^{-1} (I - \Delta F)(\tilde{M}V - \tilde{N}U) V^{-1}.
\end{aligned}$$

Because K stabilizes G, by Lemma 6.1.4, $(\tilde{M}V - \tilde{N}U)^{-1} \in \mathbf{H}_\infty$. We have that $\Delta \in \mathbf{H}_\infty$, $F \in \mathbf{H}_\infty$ and $\|\Delta F\| < 1$, so $I - \Delta F$ is invertible over $\mathbf{H}_\infty$. Therefore, all the factors in the last expression for $(I - G_\Delta K)$ have inverses in $\mathbf{H}_\infty$. We conclude that $(I - G_\Delta K)^{-1} \in \mathbf{H}_\infty(\mathcal{L}(Y))$.

Next, we prove that the closed-loop transfer function is in $\mathbf{H}_\infty$, using the expression for $I - G_\Delta K$, above:

$$
\begin{aligned}
(I - KG_\Delta)^{-1}K = K(I - G_\Delta K)^{-1} \\
= U(\tilde{M}V - \tilde{N}U)^{-1}(I - \Delta F)^{-1}(\tilde{M} + \Delta_M) \in \mathbf{H}_\infty, \\
(I - G_\Delta K)^{-1}G_\Delta = V(\tilde{M}V - \tilde{N}U)^{-1}(I - \Delta F)^{-1}(\tilde{N} + \Delta_N) \in \mathbf{H}_\infty, \\
K(I - G_\Delta K)^{-1}G_\Delta = U(\tilde{M}V - \tilde{N}U)^{-1}(I - \Delta F)^{-1}(\tilde{N} + \Delta_N) \in \mathbf{H}_\infty.
\end{aligned}
$$

Furthermore, $G_\Delta(I - KG_\Delta)^{-1}K = G_\Delta K(I - G_\Delta K)^{-1} = -I + (I - G_\Delta K)^{-1} \in \mathbf{H}_\infty$. From this, we conclude that the closed-loop transfer function given by

$$
\mathcal{T}(G_\Delta, K) = \begin{bmatrix} (I - G_\Delta K)^{-1}G_\Delta & G_\Delta(I - KG_\Delta)^{-1}K \\ K(I - G_\Delta K)^{-1}G_\Delta & (I - KG_\Delta)^{-1}K \end{bmatrix}
$$

is in $\mathbf{H}_\infty(\mathcal{L}(U \times Y, Y \times U))$.   □

To obtain a parameterization of *all* robustly stabilizing controllers, we would have to obtain equivalence in the above theorem. To do so it is necessary to construct a destabilizing perturbation. This is possible only by assuming extra smoothness conditions on the behavior of $G(s)$ on the imaginary axis (see the proof of Theorem 1 in Georgiou and Smith [48]). Because our systems do not, in general, satisfy these assumptions, their result is not relevant to our problem. As a consequence, there may be more robustly stabilizing controllers than those that are given by our parameterization.

Next, we relate the optimization problem from Theorem 7.2.2 to a Nehari extension problem. The parameterization of all solutions to the Nehari problem in Theorem 7.1.2 then will allow us to parameterize the solutions to the robust stabilization problem.

Theorem 7.2.3 *Suppose that $G \in \mathcal{M}\mathbf{H}_\infty[\mathbf{H}_\infty]^{-1}$ has a normalized doubly coprime factorization over $\mathbf{H}_\infty$. Let $0 < \varepsilon < 1$ and $K \in \mathcal{M}\mathbf{H}_\infty[\mathbf{H}_\infty]^{-1}$. The following three statements are equivalent.*

1. *K stabilizes G and*

$$
\left\| \begin{bmatrix} K \\ I \end{bmatrix} (I - GK)^{-1}\tilde{M}^{-1} \right\|_\infty \leq \frac{1}{\varepsilon}. \tag{7.30}
$$

2. *K has a right-coprime factorization $K = UV^{-1}$ over $\mathbf{H}_\infty$ satisfying*

$$
\left\| \begin{bmatrix} -\tilde{N}^\sim \\ \tilde{M}^\sim \end{bmatrix} + \begin{bmatrix} U \\ V \end{bmatrix} \right\|_\infty \leq \sqrt{1 - \varepsilon^2}. \tag{7.31}
$$

3. *There exists $U, V \in \mathcal{M}\mathbf{H}_\infty$ such that V is invertible on some right half-plane and*

$$
\left\| \begin{bmatrix} -\tilde{N}^\sim \\ \tilde{M}^\sim \end{bmatrix} + \begin{bmatrix} U \\ V \end{bmatrix} \right\|_\infty \leq \sqrt{1 - \varepsilon^2}. \tag{7.32}
$$

Proof As in the proof of the previous theorem, we use the shorthand notation

$$F(s) = \begin{bmatrix} K(s) \\ I \end{bmatrix} (I - G(s)K(s))^{-1} \tilde{M}^{-1}(s).$$

$(1 \Rightarrow 2)$. Let $K = U_1 V_1^{-1}$ be a stabilizing compensator for G, then by Lemma 6.1.4, $\tilde{M}V_1 - \tilde{N}U_1$ is invertible over $\mathbf{H}_\infty$. Consequently,

$$\begin{bmatrix} U \\ V \end{bmatrix} := -\varepsilon^2 \begin{bmatrix} U_1 \\ V_1 \end{bmatrix} (\tilde{M}V_1 - \tilde{N}U_1)^{-1}$$

is a right-coprime factorization of K if U_1, V_1 is. Because $M, N, \tilde{M}, \tilde{N}$ form a normalized doubly coprime factorization of G,

$$W(j\omega) := \begin{bmatrix} M^*(j\omega) & N^*(j\omega) \\ -\tilde{N}(j\omega) & \tilde{M}(j\omega) \end{bmatrix}$$

is co-inner, i.e., $W(j\omega)W(j\omega)^* = I$ for all $\omega \in \mathbb{R}$. As $W(j\omega)$ is a square matrix, it is also inner: $W(j\omega)^*W(j\omega) = I$ for all $\omega \in \mathbb{R}$. So, substituting $K = U_1 V_1^{-1}$ and $G = \tilde{M}^{-1}\tilde{N}$ in the formula for F, we obtain

$$\begin{aligned}
\|F\|_\infty^2 &= \|WF\|_\infty^2 \\
&= \left\| W \begin{bmatrix} U_1 \\ V_1 \end{bmatrix} (\tilde{M}V_1 - \tilde{N}U_1)^{-1} \right\|_\infty^2 \\
&= \left\| \begin{bmatrix} (M^*U_1 + N^*V_1)(\tilde{M}V_1 - \tilde{N}U_1)^{-1} \\ I \end{bmatrix} \right\|_\infty^2 \\
&= 1 + \|(M^*U_1 + N^*V_1)(\tilde{M}V_1 - \tilde{N}U_1)^{-1}\|_\infty^2 \\
&\leq \frac{1}{\varepsilon^2}.
\end{aligned}$$

Consequently,

$$\begin{aligned}
\left\| \begin{bmatrix} -\tilde{N}^* \\ \tilde{M}^* \end{bmatrix} + \begin{bmatrix} U \\ V \end{bmatrix} \right\|_\infty^2 &= \left\| W \begin{bmatrix} -\tilde{N}^* \\ \tilde{M}^* \end{bmatrix} + W \begin{bmatrix} U \\ V \end{bmatrix} \right\|_\infty^2 \\
&= \left\| \begin{bmatrix} 0 \\ I \end{bmatrix} - \varepsilon^2 \begin{bmatrix} (M^*U_1 + N^*V_1)(\tilde{M}V_1 - \tilde{N}U_1)^{-1} \\ I \end{bmatrix} \right\|_\infty^2 \\
&= (1 - \varepsilon^2)^2 + \varepsilon^4 \|(M^*U_1 + N^*V_1)(\tilde{M}V_1 - \tilde{N}U_1)^{-1}\|_\infty^2 \\
&\leq (1 - \varepsilon^2)^2 + \varepsilon^2(1 - \varepsilon^2) \\
&= 1 - \varepsilon^2.
\end{aligned}$$

$(2 \Rightarrow 1)$. Let $K = UV^{-1}$ be a right-coprime factorization of K and assume that (7.31) is satisfied. We define

$$\begin{bmatrix} E_1 \\ E_2 \end{bmatrix} = W \left(\begin{bmatrix} -\tilde{N}^* \\ \tilde{M}^* \end{bmatrix} + \begin{bmatrix} U \\ V \end{bmatrix} \right) = \begin{bmatrix} M^*U + N^*V \\ 1 + \tilde{M}V - \tilde{N}U \end{bmatrix}.$$

By (7.31), using the inner property of W, we have

$$\left\| \begin{bmatrix} E_1 \\ E_2 \end{bmatrix} \right\|^2_\infty \leq 1 - \varepsilon^2 < 1.$$

This implies that $\|E_2\|^2_\infty \leq 1 - \varepsilon^2 < 1$. Clearly, $E_2 \in \mathbf{H}_\infty(\mathcal{L}(U))$ and so $I - E_2$ is invertible over $\mathbf{H}_\infty$. By Lemma 9.4.2 in [36], $\|E_1(I - E_2)^{-1}\|^2_\infty \leq (1 - \varepsilon^2)/\varepsilon^2$. It can easily be checked that $-E_1(I - E_2)^{-1} = (M^*U + N^*V)(\tilde{M}V - \tilde{N}U)^{-1}$, which in combination with the identity

$$\|F\|^2_\infty = 1 + \|(M^*U + N^*V)(\tilde{M}V - \tilde{N}U)^{-1}\|^2_\infty,$$

which was derived in the previous part of the proof, leads to

$$\|F\|^2_\infty = 1 + \|E_1(I - E_2)^{-1}\|^2_\infty \leq 1 + \frac{1 - \varepsilon^2}{\varepsilon^2} = \frac{1}{\varepsilon^2}.$$

To show that K stabilizes G, we have to prove that $\tilde{M}V - \tilde{N}U$ is invertible over $\mathbf{H}_\infty$. This is true because $\tilde{M}V - \tilde{N}U = E_2 - I$, which is invertible over $\mathbf{H}_\infty$, as was explained above.

($2 \Rightarrow 3$). This part of the proof is trivial and therefore omitted.

($3 \Rightarrow 2$). Assume that U, V satisfy (7.32) and that V is invertible on some right half-plane. Now,

$$1 - \varepsilon^2 \geq \left\| \begin{bmatrix} -\tilde{N}^\sim \\ \tilde{M}^\sim \end{bmatrix} + \begin{bmatrix} U \\ V \end{bmatrix} \right\|^2_\infty$$

$$= \left\| W \begin{bmatrix} -\tilde{N}^* + U \\ \tilde{M}^* + V \end{bmatrix} \right\|^2_\infty$$

$$= \left\| \begin{bmatrix} M^*U + N^*V \\ I - \tilde{N}U + \tilde{M}V \end{bmatrix} \right\|^2_\infty.$$

So, $\|I - \tilde{N}U + \tilde{M}V\|^2_\infty \leq 1 - \varepsilon^2 < 1$, which implies that $Q := (\tilde{N}U - \tilde{M}V)^{-1} \in \mathbf{H}_\infty$. Consequently,

$$(Q\tilde{M})V - (Q\tilde{N})U = I$$

is a Bezout equation and U and V are right coprime. Because V is invertible on some right half-plane, $K := UV^{-1} \in \mathcal{M}\mathbf{H}_\infty[\mathbf{H}_\infty]^{-1}$.   □

The theorem above reduces the problem of robust stabilization with respect to left-coprime factor perturbations to the solution of the suboptimal Nehari problem for $\tilde{G} = \begin{bmatrix} -\tilde{N} & \tilde{M} \end{bmatrix}$, as stated in (7.32). The solution to the Nehari problem is well known (see Power [87]):

$$\inf_{J(-s) \in \mathcal{M}\mathbf{H}_\infty} \|\tilde{G} + J\|_\infty = \|H_{\tilde{G}}\|, \tag{7.33}$$

where $H_{\tilde{G}}$ is the Hankel operator with symbol $\tilde{G}$ (see Section 7.1 for the definition of a Hankel operator). The difficulty in solving this particular Nehari problem is that, unlike the case of exponential stability, the Hankel operator is not compact. In Section 7.1, a parameterization of all solutions to the suboptimal Nehari problem with noncompact Hankel operator (and assuming finite-dimensional input and output spaces) was given.

We want to solve the robust stabilization problem for systems $\Sigma(A, B, C, D)$ with finite-dimensional inputs and outputs under assumptions (i)–(vi). In this case, $\tilde{G} = [\; -\tilde{N}_P \;\; \tilde{M}_P \;]$, where $\tilde{M}_P$, $\tilde{N}_P$ are as defined in (6.20)–(6.21). Let us recall the relevant formulas,

$$\tilde{M}_P = R^{-\frac{1}{2}} + R^{-\frac{1}{2}}C(sI - A_{L_P})^{-1}L_P, \tag{7.34}$$

$$\tilde{N}_P = R^{-\frac{1}{2}}D + R^{-\frac{1}{2}}C(sI - A_{L_P})^{-1}(B + L_P D), \tag{7.35}$$

where $R = I + DD^*$, $L_P = -(PC^* + BD^*)R^{-1}$, $A_{L_P} = A + L_P C$ and P is the unique, self-adjoint, strongly stabilizing solution of the Riccati equation

$$APz + PA^*z - (PC + BD^*)R^{-1}(CP + DB^*)z + BB^*z = 0 \tag{7.36}$$

for $z \in \mathcal{D}(A^*)$. Now, $\tilde{G} = [\; -\tilde{N}_P \;\; \tilde{M}_P \;] = \tilde{D} + \tilde{C}(sI - \tilde{A})^{-1}\tilde{B}$, where

$$\tilde{A} = A_{L_P}, \quad \tilde{B} = [\; -(B + L_P D) \;\; L_P \;],$$
$$\tilde{C} = R^{-\frac{1}{2}}C, \quad \tilde{D} = [\; -R^{-\frac{1}{2}}D \;\; R^{-\frac{1}{2}} \;]. \tag{7.37}$$

We need some technical results for the controllability and observability Gramians L_B and L_C.

Lemma 7.2.4 *Let G have a normalized left-coprime factorization $G = \tilde{M}^{-1}\tilde{N}$. The Hankel operator with symbol $\tilde{G} = [\; -\tilde{N} \;\; \tilde{M} \;]$ satisfies $\|H_{\tilde{G}}\| < 1$.*

Proof Let us denote $\tilde{U} = U \times Y$ and $\tilde{Y} = Y$ (i.e., $\tilde{G}(s) : \tilde{U} \to \tilde{Y}$). Clearly, $\|H_{\tilde{G}}\| = \|H_{\tilde{G}\sim}\| \leq \|\Lambda_{\tilde{G}\sim}\| = 1$, where the first equality follows from the fact that $H_{\tilde{G}}^* = H_{\tilde{G}\sim}$ and the last equality follows from the fact that for an inner function F, $\|\Lambda_F f\| = \|f\|$. Suppose now that $\|H_{\tilde{G}\sim}\| = 1$. Then there must exist $\{f_n\}_{n\in\mathbb{N}}$ with $f_n \in \mathbf{H}_2(\tilde{Y})$, $\|f_n\| = 1$ such that $\lim_{n\to\infty} \|H_{\tilde{G}\sim} f_n\| = 1$. We define g_n, g_n^+, g_n^- by

$$g_n(j\omega) = \tilde{G}\sim(j\omega)f_n(-j\omega) \in \mathbf{L}_2(-j\infty, j\infty; \tilde{U}),$$
$$g_n^+ = \pi g_n \in \mathbf{H}_2(\tilde{U}),$$
$$g_n^- = (I - \pi)g_n \in \mathbf{H}_2^{\perp}(\tilde{U}),$$

where π is the orthogonal projection from $\mathbf{L}_2(-j\infty, j\infty; \tilde{U})$ onto $\mathbf{H}_2(\tilde{U})$. Clearly, $g_n^+ = H_{\tilde{G}\sim}f_n$ and $\|g_n^+\|_{\mathbf{H}_2}^2 + \|g_n^-\|_{\mathbf{H}_2^{\perp}}^2 = \|g_n\|_{\mathbf{L}_2}^2 = 1$, where the last equality follows from the inner property of $\tilde{G}\sim$. Thus, we have $\lim_{n\to\infty} \|g_n^+\|_{\mathbf{H}_2} = 1$ and $\lim_{n\to\infty} \|g_n^-\|_{\mathbf{H}_2^{\perp}} = 0$. For $s = -j\omega$, we have

$$\begin{bmatrix} \tilde{N}(j\omega)^* \\ \tilde{M}(j\omega)^* \end{bmatrix} f_n(j\omega) = g_n^+(-j\omega) + g_n^-(-j\omega).$$

Multiplying by $[\; -Y(j\omega)^* \;\; X(j\omega)^* \;]$, where X and Y are the Bezout factors related to the coprime factorization, we obtain

$$f_n(j\omega) = [\; -Y(j\omega)^* \;\; X(j\omega)^* \;] g_n^+(-j\omega) + [\; -Y(j\omega)^* \;\; X(j\omega)^* \;] g_n^-(-j\omega).$$

The left-hand side has an extension to a function in $\mathbf{H}_2(\tilde{Y})$, the first term on the right-hand side has an extension to a function in $\mathbf{H}_2^{\perp}(\tilde{Y})$ and the second term on the right-hand side converges to zero. This contradiction implies that $\|H_{\tilde{G}}\| \neq 1$. $\qquad\square$

Lemma 7.2.5 *Consider the system $\Sigma(A, B, C, D)$ with finite-dimensional input space and output space, under assumptions (i)–(vi). Let $\tilde{G}(s) = \begin{bmatrix} -\tilde{N}_P(s) & \tilde{M}_P(s) \end{bmatrix} = \tilde{D} + \tilde{C}(sI - \tilde{A})^{-1}\tilde{B}$. The controllability and observability Gramians of $\tilde{G}(s)$, L_B and L_C, are given by*

$$L_B = P, \tag{7.38}$$
$$L_C = Q(I + PQ)^{-1}, \tag{7.39}$$

where Q and P are the strongly stabilizing solutions of the control and filter Riccati equations (6.14), (6.15). Moreover,

$$r(L_B L_C) = r(L_C L_B) = r(PQ(I + PQ)^{-1}). \tag{7.40}$$

Proof From Lemma 7.2.4, we obtain that $r(L_B L_C) = r(L_C L_B) = \|H_{\tilde{G}}\| < 1$. Hence, $I - L_B L_C$ and $I - L_C L_B$ are boundedly invertible. The proof of Lemma 9.4.10 in [36] applies to prove that $I + PQ$ is boundedly invertible and that (7.38), (7.39) hold. The equality of the spectral radii in (7.40) is then trivial. □

Because P is a strongly stabilizing solution to the filter Riccati equation (7.36), the system $\Sigma(\tilde{A}^*, \tilde{C}^*, \tilde{B}^*, \tilde{D}^*)$ is a strongly stable system and, consequently, the dual system $\Sigma(\tilde{A}, \tilde{B}, \tilde{C}, \tilde{D})$ is input stable, output stable and input-output stable. Thus, the assumptions H1–H4 in Section 7.1 are satisfied, and we can apply the results of that section to find a parameterization of all solutions to the suboptimal Nehari extension problem for $\tilde{G}_0(s) := \tilde{C}(sI - \tilde{A})^{-1}\tilde{B}$. We will formulate the results of Theorems 7.1.1 and 7.1.2 for this special case.

Theorem 7.2.6 *Consider the system $\Sigma(A, B, C, D)$ with finite-dimensional inputs and outputs under assumptions (i)–(vi). Let $\tilde{A}, \tilde{B}, \tilde{C}, \tilde{D}$ be given by (7.37) and define $\tilde{G}_0(s) = \tilde{C}(sI - \tilde{A})^{-1}\tilde{B} = \begin{bmatrix} -\tilde{N}_P(s) & \tilde{M}_P(s) \end{bmatrix} - \tilde{D}$. The following three statements are equivalent.*

1. *There exists a $J(-s) \in \mathcal{M}\mathbf{H}_\infty$ such that*

$$\|\tilde{G}_0 + J\|_\infty < \sigma. \tag{7.41}$$

2. *$\sigma > r^{\frac{1}{2}}(L_B L_C)$, where L_B and L_C are the controllability Gramian and observability Gramian of $\Sigma(\tilde{A}, \tilde{B}, \tilde{C})$, respectively.*

3. *There exists an X satisfying the factorization*

$$\begin{bmatrix} I & \tilde{G}_0 \\ 0 & I \end{bmatrix}^{\sim} \begin{bmatrix} I & 0 \\ 0 & -\sigma^2 I \end{bmatrix} \begin{bmatrix} I & \tilde{G}_0 \\ 0 & I \end{bmatrix} = X^{\sim} \begin{bmatrix} I & 0 \\ 0 & -I \end{bmatrix} X \tag{7.42}$$

almost everywhere on the imaginary axis, which is such that $X_{11}^{-1}(-s) \in \mathcal{M}\mathbf{H}_\infty$ and

$$X(-s) - \begin{bmatrix} I & 0 \\ 0 & \sigma I \end{bmatrix} \in \mathcal{M}\mathbf{H}_2,$$
$$X^{-1}(-s) - \begin{bmatrix} I & 0 \\ 0 & \frac{1}{\sigma}I \end{bmatrix} \in \mathcal{M}\mathbf{H}_2.$$

Moreover, all $J(-s) \in \mathcal{M}\mathbf{H}_\infty$ satisfying (7.41) for $\sigma > r^{\frac{1}{2}}(L_B L_C)$ are given by $J(-s) = R_1(-s)R_2(-s)^{-1}$ with

$$\begin{bmatrix} R_1(-s) \\ R_2(-s) \end{bmatrix} = X^{-1}(-s) \begin{bmatrix} Q(-s) \\ I \end{bmatrix}$$

for some $Q(-s) \in \mathcal{M}\mathbf{H}_\infty$ satisfying $\|Q\|_\infty \leq 1$. An X satisfying (7.42) and its inverse are given by

$$X(s) = \begin{bmatrix} I & 0 \\ 0 & \sigma I \end{bmatrix} + \frac{1}{\sigma^2} \begin{bmatrix} -\tilde{C}L_B \\ \sigma \tilde{B}^* \end{bmatrix} N_\sigma^* (sI + \tilde{A}^*)^{-1} \begin{bmatrix} \tilde{C}^* & L_C \tilde{B} \end{bmatrix}, \quad (7.43)$$

$$X^{-1}(s) = \begin{bmatrix} I & 0 \\ 0 & \frac{1}{\sigma}I \end{bmatrix} + \frac{1}{\sigma^2} \begin{bmatrix} \tilde{C}L_B \\ -\tilde{B}^* \end{bmatrix} (sI + \tilde{A}^*)^{-1} N_\sigma^* \begin{bmatrix} \tilde{C}^* & \frac{1}{\sigma}L_C\tilde{B} \end{bmatrix}, \quad (7.44)$$

where $N_\sigma = (I - \frac{1}{\sigma^2}L_B L_C)^{-1}$.

Now we can proceed as in Chapter 9 of [36] to obtain a parameterization of stabilizing controllers for $G(s)$.

Theorem 7.2.7 *Consider the robust stabilization problem for $\Sigma(A, B, C, D)$ with finite-dimensional inputs and outputs under assumptions (i)–(vi).*

1. *The maximum robustness margin is $\varepsilon_{max} = \sqrt{1 - r(PQ(I + PQ)^{-1})}$.*

2. *All controllers $K \in \mathcal{M}\mathbf{H}_\infty[\mathbf{H}_\infty]^{-1}$ given by*

$$K(s) = [\phi_{11}(s)L(s) + \phi_{12}(s)][\phi_{21}(s)L(s) + \phi_{22}(s)]^{-1} \quad (7.45)$$

for arbitrary $L \in \mathbf{H}_\infty$ satisfying $\|L\|_\infty \leq 1$ are robustly stabilizing for $G(s) = D + C(sI - A)^{-1}B$, achieving a robustness margin $0 < \varepsilon < \varepsilon_{max}$, where

$$\begin{bmatrix} \phi_{11} & \phi_{12} \\ \phi_{21} & \phi_{22} \end{bmatrix} = \begin{bmatrix} \frac{1}{\varepsilon}S^{-\frac{1}{2}} & \frac{1}{\sigma}D^*R^{-\frac{1}{2}} \\ \frac{1}{\varepsilon}DS^{-\frac{1}{2}} & -\frac{1}{\sigma}R^{-\frac{1}{2}} \end{bmatrix} \quad (7.46)$$

$$+ \frac{1}{\sigma^2}\begin{bmatrix} F_Q \\ C + DF_Q \end{bmatrix}(sI - A_{F_Q})^{-1}W^*\begin{bmatrix} \frac{1}{\varepsilon}BS^{-\frac{1}{2}} & -\frac{1}{\sigma}PC^*R^{-\frac{1}{2}} \end{bmatrix},$$

$W = (I - \frac{1}{\sigma^2}QP + QP)^{-1}$, $S = I + D^*D$, $R = (I + DD^*)$, $F_Q = -S^{-1}(DB^* + B^*Q)$, $\sigma = \sqrt{1 - \varepsilon^2}$ and P and Q are the unique, self-adjoint, strongly stabilizing solutions of the control and filter Riccati equations (6.14), (6.15), respectively.

Proof The proof is a matter of manipulating the formulas and is a straightforward extension of the proofs of Lemmas 9.4.13 and 9.4.14 and Theorem 9.4.15 in [36]. □

A popular choice in the literature is to use the *central controller*, which amounts to the choice $L(s) \equiv 0$ in (7.45).

Corollary 7.2.8 *Consider the system $\Sigma(A, B, C, D)$ under the assumptions of Theorem 7.2.7. Let P, Q, F_Q and W be as in the previous theorem. The controller*

$$K_0(s) = -D - \frac{1}{\sigma^2}B^*Q(sI - A_0)^{-1}W^*PC^*,$$

where $\sigma^2 = 1 - \varepsilon^2$ and

$$A_0 = A + BF_Q - \frac{1}{\sigma^2}W^*PC^*(C + DF_Q),$$

stabilizes the system with a robustness margin of ε, $0 < \varepsilon < \varepsilon_{max}$.

Finally, we take a look at the important special case of systems in the $\Sigma(A, B, B^*)$ class satisfying $A^* = -A$ and $D = 0$. The examples in Sections 9.1 and 9.2 have this special form, and the example in Section 9.3 has this form if its parameters satisfy $d_0 = d_1 = 0$.

Corollary 7.2.9 *Consider a system $\Sigma(A, B, B^*)$ satisfying A1–A5 of Section 2.2 and with the additional assumptions that U is finite-dimensional and $A + A^* = 0$. The robustness margin satisfies*

$$\varepsilon_{max} = \frac{1}{\sqrt{2}}$$

and all controllers given by

$$K(s) = [\phi_{11}(s)L(s) + \phi_{12}(s)][\phi_{21}(s)L(s) + \phi_{22}(s)]^{-1}$$

are robustly stabilizing controllers for $G(s) = B^*(sI - A)^{-1}B$, with robustness margin $0 < \varepsilon < \varepsilon_{max}$, where

$$\begin{bmatrix} \phi_{11}(s) & \phi_{12}(s) \\ \phi_{21}(s) & \phi_{22}(s) \end{bmatrix} = \begin{bmatrix} \frac{1}{\varepsilon}I & 0 \\ 0 & -\frac{1}{\sigma}I \end{bmatrix}$$
$$+ \frac{1}{2\sigma^2 - 1}\begin{bmatrix} -I \\ I \end{bmatrix} B^*(sI - A + BB^*)^{-1}B \begin{bmatrix} \frac{1}{\varepsilon}I & -\frac{1}{\sigma}I \end{bmatrix},$$

$\sigma = \sqrt{1 - \varepsilon^2}$ and $L \in \mathcal{M}\mathbf{H}_\infty$ satisfies $\|L\|_\infty \leq 1$.

In particular, the controller $K = -I$, which corresponds to $L(s) = \varepsilon/\sqrt{1 - \varepsilon^2}I$, is robustly stabilizing with the maximal robustness margin of $1/\sqrt{2}$. Furthermore, the central controller is given by

$$K_0(s) = \frac{-1}{1 - \varepsilon^2}B^*(sI - A_0)^{-1}B,$$

where $A_0 = A - (2 - \varepsilon^2)/(1 - \varepsilon^2)BB^*$.

Proof It is easy to see that with $D = 0$ and $A + A^* = 0$, the solutions of the Riccati equations must satisfy $P = Q = I$. All the results then follow directly from Theorem 7.2.7. That $K = -I$ stabilizes with the maximal robustness margin was also proven in Curtain and van Keulen [33]. □

The most important class of systems satisfying conditions (i)–(vi) is that of strongly stabilizable and detectable systems $\Sigma(A, B, C, D)$, where A has compact resolvent. For these systems, by Theorem 6.3.3, the robustly stabilizing controllers $K(s)$ in Theorem 7.2.7 yield a strongly stable closed-loop system if $K(s)$ has a strongly stabilizable and strongly detectable realization. In particular, the central controller $K_0(s)$ for systems $\Sigma(A, B, B^*)$ with $A + A^* = 0$ has a statically stabilizable (even strongly stable) realization $\Sigma(A - \frac{2-\varepsilon^2}{1-\varepsilon^2}BB^*, B, \frac{-1}{1-\varepsilon^2}B^*, 0)$, from Corollary 7.2.9. Therefore, this controller will strongly stabilize the systems $\Sigma(A, B, B^*)$ with $A + A^* = 0$.

Chapter 8

Robustness with Respect to Nonlinear Perturbations

A common way to stabilize a linear system is by the use of linear feedback, but in implementing this linear feedback, nonlinear effects may be introduced. The question then arises whether these nonlinearities could render the closed-loop system unstable. In this chapter, we examine this question for strongly stabilizable infinite-dimensional systems. In Chapter 9, we present examples of models of engineering systems that are not exponentially stabilizable, but merely strongly stabilizable. These examples show that there is a need to extend results concerning robustness of asymptotic stability in the presence of nonlinear perturbations to the case of strongly stabilizable systems. This is especially the case because many of the actuators and sensors in structural control are based on so-called smart materials, for instance, piezoelectric and magnetostrictive materials, which show strong nonlinear effects such as saturations, hysteresis and deadzones.

If a system is strongly stable but not exponentially stable, then it will have poles arbitrarily close to the imaginary axis. In this sense, there is no stability margin. This may lead one to think that strong stability is a very nonrobust property, but this is not the case. In Chapter 7, we have shown that robustness with respect to *linear* coprime factor perturbations can be obtained. In this chapter, we prove robustness with respect to certain classes of static and dynamic *nonlinear* perturbations.

A classical mathematical formulation of the problem of robustness of asymptotic stability in the presence of nonlinear perturbations is the *absolute stability problem:* Under what conditions on the stable system Σ will the feedback connection as in Figure 8.1 of the system with a nonlinearity remain stable for all functions ϕ in a given class of nonlinearities?

We present one famous criterion of absolute stability. We prove a version of the celebrated *Popov criterion* for a class of strongly stable systems in Section 8.2. As an application of this criterion, we derive a result about asymptotic tracking of constant reference signals using integral control. However, before presenting these robustness results, we introduce in Section 8.1 the necessary concepts related to nonlinear differential equations on Hilbert spaces.

Another interesting result in the literature that can be interpreted as a type of robustness of asymptotic stability with respect to nonlinear perturbations is from Slemrod [94], who showed that the feedback $u = -y$ strongly stabilizes a system in the $\Sigma(A, B, B^*)$ class, even in the presence of saturation effects. This very nice result, which is true only for a

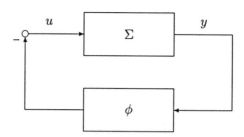

Figure 8.1: Feedback interconnection of linear system Σ and nonlinearity ϕ.

scalar control, is not included in our results in this chapter, nor does it include our results: they complement one another.

8.1   Preliminaries

We give some definitions concerning various solution concepts and stability concepts for a nonlinear differential equation on a Hilbert space Z. We use the following solution concepts, which have been taken from Pazy [85, Chapter 4, Definitions 2.3 and 2.8].

Definition 8.1.1   *Let A generate a C_0-semigroup $T(t)$ on Z. A function $z(t) \in C(0, t_1; Z)$ is a* mild solution *of the differential equation*

$$\dot{z}(t) = Az(t) + f(z(t), t), \quad z(0) = z_0, \tag{8.1}$$

on $[0, t_1]$ if it satisfies the integral equation

$$z(t) = T(t)z_0 + \int_0^t T(t-s)f(z(s), s)ds.$$

If a mild solution is differentiable almost everywhere, $\dot{z}(t) \in \mathbf{L}_1(0, t_1; Z)$, and it satisfies (8.1) almost everywhere in $(0, t_1)$, then it is called a strong solution.

There are many different possible notions of stability for nonlinear differential equations on Hilbert spaces. Here, we define the types of stability in which we will be interested. The terminology was taken from the book by Willems [119]. The precise formulation for the infinite-dimensional case is as in Wexler [115, 116], who studied the Popov criterion for a class of exponentially stable infinite-dimensional systems occurring as models of a nuclear reactor.

Definition 8.1.2   *Consider the nonlinear differential equation (8.1). Assume that it has the origin as an equilibrium point, i.e., $f(0, t) = 0$ for all $t \geq 0$. If*

(i) for any $z_0 \in Z$, (8.1) has a unique mild solution $z(t)$ defined for $t \in [0, \infty)$, and

(ii) there exists a strictly increasing continuous function $\pi : \mathbb{R}^+ \to \mathbb{R}^+$ with $\pi(0) = 0$ such that for any solution $z(t)$ to (8.1) with initial condition z_0 and any $r > 0$,

$$\|z_0\| \leq r \implies \|z(t)\| \leq \pi(r),$$

then the system is called stable in the large. *If, in addition,*

(iii) for any $z_0 \in Z$ the unique mild solution $z(t) \to 0$ as $t \to \infty$,

then we call the system globally asymptotically stable. *The system is called* uniformly asymptotically stable in the large *if the conditions (i), (ii) and (iii) hold and also*

(iv) *for any z_0 in a bounded set $\mathcal{B} \in Z$, each solution $z(t) \to 0$ as $t \to \infty$ uniformly with respect to $z_0 \in \mathcal{B}$.*

8.2 The Popov Criterion

In this section, we prove a version of the Popov criterion for strongly stable infinite-dimensional systems. This criterion is concerned with the asymptotic stability of a stable linear system connected in feedback with an integrator and a nonlinear function ϕ. It states that a sufficient condition for the closed-loop system to remain asymptotically stable for all ϕ in the sector $[0, \infty)$ is that the transfer function of the linear system is strictly positive real.

This criterion was derived by Popov, and his article [86] was the starting point for many other papers. Narendra and Taylor devoted a large part of their monograph [79] to it. A short and very accessible explanation of the Popov criterion can be found in Vidyasagar [105]. In Wexler [115, 116] and Bucci [20], extensions of the Popov criterion to various classes of infinite-dimensional systems were developed, but all these results had the crucial assumption that the linear system is exponentially stable.

The nonlinearities under consideration in this problem are so-called sectorial nonlinearities, i.e., their graphs are contained in a cone. A function $\phi : \mathbb{R} \to \mathbb{R}$ is said to be in the sector $[\alpha, \beta]$ if $\alpha x^2 \leq x\phi(x) \leq \beta x^2$. In particular, a function is in the sector $[0, \infty)$ if it satisfies $y\phi(y) \geq 0$. In that case, its graph is contained in the first and third quadrants. This class of nonlinearities includes two of the most important nonlinearities occurring in practice: saturations and deadzones. To illustrate the nonlinearities for which stability is obtained, we give two graphs of typical nonlinearities. The function in the left-hand graph in Figure 8.2 satisfies $y\phi(y) \geq 0$. For this kind of nonlinearity, we obtain stability in the large and convergence of $z(t)$ to zero. The right-hand one is a graph of a function satisfying $y\phi(y) > 0$ for $y \neq 0$. It is for this type of nonlinearity that we obtain uniform asymptotic stability in the large.

Next, we formulate our version of the Popov criterion.

Theorem 8.2.1 *Consider the strongly stable, single-input, single-output system*

$$\begin{aligned}
\dot{z}(t) &= Az(t) + Bu(t), \quad z(0) = z_0, \\
y(t) &= Cz(t) + du(t),
\end{aligned} \tag{8.2}$$

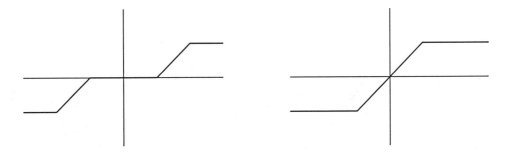

Figure 8.2: Graphs of typical nonlinearities.

with transfer function $G(s)$, in feedback connection with the nonlinearity

$$\dot{\sigma}(t) = y(t), \quad \sigma(0) = \sigma_0,$$
$$u(t) = -\phi(\sigma(t)). \tag{8.3}$$

Let $\phi : \mathbb{R} \to \mathbb{R}$ be any function satisfying the following conditions:

$$y\phi(y) \geq 0 \text{ for all } y \in \mathbb{R}, \tag{8.4}$$

$$\lim_{|\sigma| \to \infty} \int_0^\sigma \phi(s)ds = +\infty, \tag{8.5}$$

$$|\phi(r_1) - \phi(r_2)| \leq l(r)|r_1 - r_2| \text{ if } |r_1| \leq r, |r_2| \leq r \tag{8.6}$$

for some strictly increasing continuous function $l(r)$. Furthermore, assume that $G(s)$ is strictly positive real. Then, the system is stable in the large and $z(t) \to 0$ as $t \to \infty$. If, in addition, $y\phi(y) > 0$ for $y \neq 0$, then the system is uniformly asymptotically stable in the large.

The proof of this result is given in Section 8.2.1. It uses the fact that if $\Sigma(A, B, C, d)$ is strictly positive real, then $d > 0$ and there exists an $\varepsilon \in (0, d)$ such that $\Sigma(A, B, C, \varepsilon)$ is also strictly positive real. From the strict positive real lemma (see Section 4.5.2), it follows that there exists a strongly stabilizing solution $P = P^* \geq 0$ to the following Riccati equation for $z \in \mathcal{D}(A)$:

$$A^*Pz + PAz + \frac{1}{2\varepsilon}(PB - C^*)(B^*P - C)z = 0. \tag{8.7}$$

The approach is similar to that in Wexler [116] and Bucci [20]. However, in those papers the exponential stability property of the semigroup is used in a crucial manner. Since we do not have this property, we need to modify the proofs at several steps.

Let us comment on the class of nonlinearities. By (8.4), the nonlinearities are in the sector $[0, \infty)$, i.e., they have graphs lying in the first and third quadrants. By continuity, this implies that $\phi(0) = 0$. Equation (8.6) is a local Lipschitz condition. This condition is necessary to ensure the existence of a strong solution. As mentioned in the introduction, this class of nonlinearities does include saturation effects and deadzones.

For the special case of systems of the form $\Sigma(A_B, B, B^*, d)$ under assumptions A1–A5 of Section 2.2 with $d > 0$ and where $A_B = A - BB^*$, the strong stability and strict positive real property are automatically satisfied by properties P1–P6 of Lemma 2.2.6. Therefore, for these systems we obtain the following corollary.

Corollary 8.2.2 *Consider the system $\Sigma(A_B, B, B^*, d)$ under assumptions A1–A5 with $U = \mathbb{R}$ and $d > 0$ in connection with a nonlinearity ϕ as in (8.3). If $\phi : \mathbb{R} \to \mathbb{R}$ satisfies (8.4)–(8.6), then the system is stable in the large and $z(t) \to 0$ as $t \to \infty$. Under the additional assumption that $y\phi(y) > 0$ for $y \neq 0$, the system is uniformly asymptotically stable in the large.*

In Section 8.2.2, we apply the Popov criterion to show that integral control can be used to track constant reference signals for strongly stable distributed parameter systems in the presence of sectorial nonlinearities.

Balakrishnan [6] has derived a Popov-like criterion for global asymptotic stability of a flexible beam in the presence of sectorial nonlinearities in the feedback loop.

8.2.1 Proof of the Popov Criterion

We prove this result in a sequence of lemmas, following the general approach in Bucci [20].

Lemma 8.2.3 *The system (8.2)–(8.3) has a unique strong solution* $(z(t),\sigma(t))$ *on the interval* $[0,t_{max}(z_0,\sigma_0))$ *for some* $t_{max}(z_0,\sigma_0) \leq \infty$. *If* $t_{max}(z_0,\sigma_0) < \infty$, *then*

$$\lim_{t \uparrow t_{max}(z_0,\sigma_0)} \|(z(t),\sigma(t))\| = \infty.$$

Moreover, $\sigma \in C^1([0,t_{max}(z_0,\sigma_0)))$.

Proof We formulate the extended system

$$\frac{d}{dt}\begin{pmatrix} z \\ \sigma \end{pmatrix} = \begin{bmatrix} A & 0 \\ C & 0 \end{bmatrix}\begin{pmatrix} z \\ \sigma \end{pmatrix} - \begin{bmatrix} B \\ d \end{bmatrix}\phi(\sigma) \tag{8.8}$$

and note that this system satisfies the assumptions of Theorem 1.4 of Pazy [85, p. 185]. This theorem shows that (8.8) has a unique mild solution on $[0,t_{max}(z_0,\sigma_0))$ and, furthermore, that

$$\lim_{t \uparrow t_{max}(z_0,\sigma_0)} \|(z(t),\sigma(t))\| = \infty \quad \text{if } t_{max}(z_0,\sigma_0) < \infty.$$

By the continuity of σ and z and the Lipschitz continuity of ϕ, it follows from the expression

$$\sigma(t) = \sigma_0 + \int_0^t Cz(s)ds - d\int_0^t \phi(\sigma(s))ds$$

that $\sigma \in C^1([0,T])$ for any $T < t_{max}(z_0,\sigma_0)$. Hence, $\phi(\sigma)$ is Lipschitz on $[0,T]$ and Corollary 2.11 of Pazy [85, p. 109] applies to show that, in fact, we have a strong solution. The solution on $[0,t_{max}(z_0,\sigma_0))$ is given by

$$z(t) = T(t)z_0 - \int_0^t T(t-s)B\phi(\sigma(s))ds,$$

$$\sigma(t) = \sigma_0 + \int_0^t Cz(s)ds - d\int \phi(\sigma(s))ds,$$

and so

$$\dot{\sigma}(t) = Cz(t) - d\phi(\sigma(t)).$$

Since $(z(t),\sigma(t))$ is a mild solution, it is continuous. Consequently, $\phi(\sigma(\cdot))$ is continuous as well. Hence, $\sigma \in C^1([0,t_{max}(z_0,\sigma_0)))$. $\square$

Lemma 8.2.4 *If* $\|(z_0,\sigma_0)\| \leq r$, *then*

$$\int_0^{\sigma(t)} \phi(s)ds \leq \frac{r^2}{2}(\|P\| + l(r)) \tag{8.9}$$

for $0 \leq t \leq t_{max}(z_0,\sigma_0)$.

Proof Let

$$V(z,\sigma) = \langle Pz,z \rangle + 2\int_0^\sigma \phi(s)ds, \tag{8.10}$$

where P is the nonnegative solution of (8.7). Note that $V(0,0) = 0$ and $V(z,\sigma) \geq 0$. Consider (8.8) for initial conditions $z(0) = z_0$, $\sigma(0) = \sigma_0$. We now differentiate (8.10) along the trajectories of (8.8), assuming first that $z_0 \in \mathcal{D}(A)$. Then, for almost all $t \in [0, t_{max}(z_0, \sigma_0))$, there holds

$$
\begin{aligned}
\frac{dV}{dt} &= \langle P\dot{z}, z \rangle + \langle Pz, \dot{z} \rangle + 2\phi(\sigma)\dot{\sigma} \\
&= \langle (A^*P + PA)z, z \rangle - 2\phi(\sigma)(B^*Pz - Cz + d\phi(\sigma)) \\
&= -\frac{1}{2\varepsilon} \|(B^*P - C)z\|^2 - 2\phi(\sigma)(B^*Pz - Cz + d\phi(\sigma)) \\
&= -\left(\frac{1}{\sqrt{2\varepsilon}}(B^*P - C)z + \sqrt{2\varepsilon}\phi(\sigma) \right)^2 - 2(d - \varepsilon)\phi(\sigma)^2 \\
&\leq -2(d - \varepsilon)\phi(\sigma)^2.
\end{aligned}
$$

Note that the above result is nonpositive because we took $\varepsilon < d$. So, for $z_0 \in \mathcal{D}(A)$ and almost all $t \in [0, t_{max}(z_0, \sigma_0))$, we have

$$
\frac{d}{dt}\left[\langle Pz(t), z(t) \rangle + 2\int_0^{\sigma(t)} \phi(s)ds \right] \leq -2(d - \varepsilon)\phi(\sigma)^2 \leq 0
$$

and integrating from 0 to t gives

$$
\begin{aligned}
\langle Pz(t), z(t) \rangle + 2\int_0^{\sigma(t)} \phi(s)ds &+ 2(d - \varepsilon)\int_0^t \phi(\sigma(s))^2 ds \qquad (8.11) \\
&\leq \langle Pz_0, z_0 \rangle + 2\int_0^{\sigma_0} \phi(s)ds.
\end{aligned}
$$

This extends to all $t \in [0, t_{max}(z_0, \sigma_0))$ and $z_0 \in Z$. Note that (8.4) implies that both for $\sigma < 0$ and $\sigma > 0$, we have

$$
\int_0^{\sigma} \phi(s)ds > 0.
$$

So, if $0 < -\sigma = a \leq r$,

$$
0 < \int_0^{-a} \phi(s)ds = -\int_0^a \phi(-s)ds \leq \int_0^a l(r)s\,ds = \frac{1}{2}a^2 l(r) \leq \frac{1}{2}r^2 l(r),
$$

where we have used $\phi(0) = 0$ and (8.6). Similarly, for $r \geq \sigma > 0$,

$$
\int_0^{\sigma} \phi(s)ds \leq \frac{1}{2}r^2 l(r).
$$

Thus, (8.11) implies that for $0 \leq t < t_{max}(z_0, \sigma_0)$ and $\|(z_0, \sigma_0)\| \leq r$,

$$
\begin{aligned}
\langle Pz(t), z(t) \rangle + 2\int_0^{\sigma(t)} \phi(s)ds &\leq \|P\|\|z_0\|^2 + 2\int_0^{\sigma_0} \phi(s)ds \\
&\leq r^2(\|P\| + l(r)).
\end{aligned}
$$

This implies (8.9).   □

   Lemma 8.2.5 *The system (8.2)–(8.3) has a unique strong solution on $[0, \infty)$. Furthermore, there exists a strictly increasing, continuous function $\pi(\cdot) : \mathbb{R}^+ \to \mathbb{R}^+$ with $\pi(0) = 0$ such that $\|(z(t), \sigma(t))\| \leq \pi(r)$ if $\|(z_0, \sigma_0)\| \leq r$.*

Proof Suppose that $\|(z_0, \sigma_0)\| \le r$. Let $[0, t_{max}(z_0, \sigma_0))$ be the interval on which (8.2)–(8.3) has a unique strong solution $(z(t), \sigma(t))$. We want to prove that $\|z(t)\|$ and $|\sigma(t)|$ are bounded. To show that $|\sigma(t)|$ is bounded, we define the function $g(r) : \mathbb{R} \to \mathbb{R}^+$ by

$$g(r) = \int_0^r \phi(s) ds.$$

Then, g is nonincreasing on $\mathbb{R}^-$, g is nondecreasing on $\mathbb{R}^+$, g is continuous and

$$\begin{aligned}
\lim_{r \to -\infty} g(r) &= +\infty, \\
\lim_{r \to +\infty} g(r) &= +\infty.
\end{aligned} \tag{8.12}$$

Note that from (8.12) it follows that for $\|(\sigma_0, z_0)\| \le r$,

$$g(\sigma(t)) \le \frac{r^2}{2}(\|P\| + l(r)). \tag{8.13}$$

Combining (8.12) and (8.13) with the fact that g is continuous on $\mathbb{R}$, nonincreasing on $\mathbb{R}^-$ and nondecreasing on $\mathbb{R}^+$, we can conclude that there must exist a continuous, nondecreasing function $\eta : \mathbb{R}^+ \to \mathbb{R}^+$ such that for $\sigma \in \mathbb{R}$,

$$g(\sigma) \le \frac{r^2}{2}(\|P\| + l(r)) \Rightarrow |\sigma| \le \eta(r).$$

Hence, if $\|(z_0, \sigma_0)\| \le r$, then for all $t \ge 0$,

$$|\sigma(t)| \le \eta(r). \tag{8.14}$$

Next, we want to find a similar estimate for $\|z(t)\|$. First, we derive an estimate for the $\mathbf{L}_2$-norm of $\phi(\sigma(\cdot))$. From (8.11), we have

$$\begin{aligned}
\|\phi(\sigma(\cdot))\|_{L_2}^2 &= \int_0^\infty \phi(\sigma(s))^2 ds \\
&\le \frac{1}{2(d - \varepsilon)} \langle P z_0, z_0 \rangle + \frac{1}{d - \varepsilon} \int_0^{\sigma_0} \phi(s) ds \\
&\le \frac{r^2}{2(d - \varepsilon)}(\|P\| + l(r)),
\end{aligned}$$

where we have used (8.12). Now,

$$\begin{aligned}
\|z(t)\| &\le \|T(t) z_0\| + \left\| \int_0^t T(t - s) B \phi(\sigma(s)) ds \right\| \\
&\le \|T(t)\| \|z_0\| + \|\Phi_t\| \|\phi(\sigma(\cdot))\|_{L_2} \\
&\le \left(M + \|\tilde{\Phi}\| \frac{1}{\sqrt{2(d - \varepsilon)}}(\|P\| + l(r))^{\frac{1}{2}} \right) r \\
&=: \mu(r),
\end{aligned} \tag{8.15}$$

where $M = \sup_{t \in [0, \infty)} \|T(t)\|$ is finite by the strong stability of $T(t)$ and $\tilde{\Phi}$ is the extended input map of $\Sigma(A, B, C)$, which is bounded by the strong stability of this system.

Combining the bounds for $|\sigma(t)|$ and $\|z(t)\|$, we obtain for $t \in [0, t_{max}(z_0, \sigma_0))$,

$$\|(z(t), \sigma(t))\| \leq \sqrt{\mu(r)^2 + \eta(r)^2}$$
$$=: \pi(r),$$

where $\pi : \mathbb{R}^+ \to \mathbb{R}^+$ is a continuous, strictly increasing function with $\pi(0) = 0$. It now follows from Lemma 8.2.3 that $t_{max}(z_0, \sigma_0) = \infty$. □

Proof of Theorem 8.2.1 First, properties (i)–(ii) in Definition 8.1.2 were proved in Lemma 8.2.5. Next, we show the convergence of $z(\cdot)$ and $\phi(\sigma(\cdot))$. In the proof of Lemma 8.2.5 it was shown that $\phi(\sigma(\cdot)) \in \mathbf{L}_2(0, \infty)$. From (8.8), we have

$$z(t) = T(t)z_0 - \int_0^t T(t-s)B\phi(\sigma(s))ds.$$

The fact that $\phi(\sigma(\cdot)) \in \mathbf{L}_2(0, \infty; U)$ together with the strong stability of $T(t)$ and the input stability of $\Sigma(A, B, C)$ shows that $z(t) \to 0$ as $t \to \infty$ by Lemma 2.1.3.

Next, we show that $h(t) = |\phi(\sigma(t))|^2$ is uniformly continuous on $\mathbb{R}^+$. Recall from the proof of Lemma 8.2.5 that if $\|(z_0, \sigma_0)\| \leq r$, then $|\sigma(t)| \leq \eta(r)$. Thus,

$$
\begin{aligned}
|h(t) - h(s)| &\leq |\phi(\sigma(t)) - \phi(\sigma(s))| \cdot |\phi(\sigma(t)) + \phi(\sigma(s))| \\
&\leq l(\eta(r))|\sigma(t) - \sigma(s)| \cdot (|\phi(\sigma(t))| + |\phi(\sigma(s))|) \\
&\leq 2l(\eta(r))^2|\sigma(t) - \sigma(s)|\eta(r),
\end{aligned}
$$

where the two last inequalities hold by (8.14) and (8.6). In Lemma 8.2.3, we showed that σ is in $C^1([0, t_{max}(z_0, \sigma_0)))$ and

$$|\sigma(t) - \sigma(s)| = \left| \int_s^t \dot{\sigma}(\tau)d\tau \right| \leq |t - s| \sup_{\tau \in [0, t_{max}(z_0, \sigma_0))} |\dot{\sigma}(\tau)|.$$

Now, from (8.9) we obtain

$$
\begin{aligned}
\sup_{t \geq 0} |\dot{\sigma}(t)| &\leq \sup_{t \geq 0} |Cz(t)| + d|\phi(\sigma(t))| \\
&\leq \|c\|\mu(r) + dl(\eta(r))\eta(r) \quad \text{if } \|(z_0, \sigma_0)\| \leq r
\end{aligned}
$$

by (8.6), (8.14) and (8.15). So, $h(t)$ is uniformly continuous and with (8.15) we can apply Barbalat's lemma (see Corduneanu [22]) to obtain

$$\lim_{t \to \infty} \phi(\sigma(t)) = 0,$$

which implies that σ tends to $\phi^{-1}(0)$.

In the case that $y\phi(y) > 0$ for $y \neq 0$, $\phi^{-1}(0) = \{0\}$, which implies that $\sigma(t) \to 0$ as $t \to \infty$. Property (iv) can be shown as in Wexler [115, Theorem 1, Step II]. □

8.2.2 Asymptotic Tracking

In this section, we apply the Popov criterion (Theorem 8.2.1) to prove a result about asymptotic tracking of constant reference signals by integral control, in the presence of monotonic nonlinearities in the loop. There is an enormous amount of literature about tracking by integral control and robustness of tracking with respect to nonlinear perturbations. For instance, in the infinite-dimensional case, Logemann, Ryan and Townley [70] showed that if $G(s)$ is the transfer function of a single-input, single-output, *exponentially stable* linear system and $G(0) > 0$, then using low gain integral control means that the output $y(t)$ converges to the constant reference signal r as $t \to \infty$ for all nondecreasing locally Lipschitz nonlinearities ϕ, provided that ϕ satisfies a certain compatibility condition. The setting of the problem can be represented schematically as in Figure 8.3.

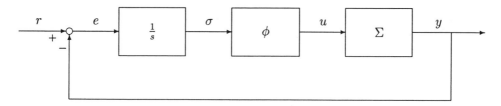

Figure 8.3: Configuration for tracking.

We prove an asymptotic tracking result like that in [70] for *strongly stable* systems $\Sigma(A, B, C, D)$ with strictly positive real transfer function $G(s)$. The nonlinearities $\phi : \mathbb{R} \to \mathbb{R}$ that we allow are nondecreasing and locally Lipschitz; i.e., there exists an $l(r) > 0$ such that for all $\sigma_1, \sigma_2 \in \mathbb{R}$, with absolute value less than r,

$$|\phi(\sigma_1) - \phi(\sigma_2)| \leq l(r)|\sigma_1 - \sigma_2|. \tag{8.16}$$

These nonlinearities include very important perturbations such as saturations and deadzones.

Theorem 8.2.6 *Consider the block diagram of Figure 8.3 under the following assumptions:*

1. *$\Sigma(A, B, C, D)$ is strongly stable and its transfer function $G(s)$ is strictly positive real;*

2. *$\phi : \mathbb{R} \to \mathbb{R}$ is locally Lipschitz and nondecreasing;*

3. *A is invertible and r is constant, so $\phi_r := G(0)^{-1}r$ is well defined;*

4. *there exists a $\sigma \in \mathbb{R}$ such that $\phi(\sigma) = \phi_r$.*

Then, the output $y(t)$ of the controlled system in Figure 8.3 converges to r as $t \to \infty$.
Proof The system is given by

$$\begin{aligned}
\dot{z}(t) &= Az(t) + Bu(t), & z(0) &= z_0, \\
\dot{\sigma}(t) &= r - y(t), & \sigma(0) &= \sigma_0, \\
y(t) &= Cz(t) + Du(t), \\
u(t) &= \phi(\sigma(t)).
\end{aligned}$$

We transform the problem in two steps into the problem of the Popov criterion. First, we transform

$$\begin{aligned}
\tilde{\sigma} &:= \sigma_0 - \sigma(t), \\
\psi_1(\tilde{\sigma}) &:= -\phi(\sigma_0 - \tilde{\sigma}).
\end{aligned}$$

Now, the system reads

$$
\begin{aligned}
\dot{z}(t) &= Az(t) + Bu(t), & z(0) &= z_0, \\
\dot{\tilde{\sigma}}(t) &= y(t) - r, & \tilde{\sigma}(0) &= 0, \\
y(t) &= Cz(t) + Du(t), & \\
u(t) &= -\psi_1(\tilde{\sigma}(t)).
\end{aligned}
$$

Condition 4 of the theorem ensures the existence of a number $\bar{\sigma}$ such that $\psi_1(\bar{\sigma}) = \psi_r :=$ $-\phi_r = -G(0)^{-1}r$, so we may introduce the variables

$$
\begin{aligned}
z_\Delta(t) &:= z(t) - A^{-1}B\psi_r, \\
\sigma_\Delta(t) &:= \tilde{\sigma} - \bar{\sigma}, \\
y_\Delta(t) &:= y(t) - r, \\
u_\Delta(t) &:= u(t) + \psi_r.
\end{aligned}
$$

Moreover, we introduce the new function ψ given by

$$
\psi(\sigma_\Delta) := \psi_1(\sigma_\Delta + \bar{\sigma}) - \psi_r.
$$

Thus, we transformed the tracking problem into the problem of asymptotic stability of

$$
\begin{aligned}
\dot{z}_\Delta(t) &= Az_\Delta(t) + Bu_\Delta(t), & z_\Delta(0) &= z_0 - A^{-1}B\psi_r, \\
\dot{\sigma}_\Delta(t) &= y_\Delta(t), & \sigma_\Delta(0) &= -\bar{\sigma}, \\
y_\Delta(t) &= Cz_\Delta(t) + Du_\Delta(t), & \\
u_\Delta(t) &= -\psi(\sigma_\Delta(t)).
\end{aligned}
$$

The condition that ϕ is locally Lipschitz and nondecreasing implies that ψ is also locally Lipschitz and nondecreasing and, hence, ψ satisfies a condition like (8.6). By construction, $\psi(0) = 0$ and $y\psi(y) \geq 0$ for all $y \in \mathbb{R}$. We can now apply Theorem 8.2.1 to our transformed delta system to show that for t tending to infinity, $\psi(\sigma_\Delta(t)) \to 0$ and $z_\Delta(t) \to 0$, which implies that $y_\Delta(t) \to 0$ and so $y(t) \to r$ as $t \to \infty$. $\quad\square$

If A generates a strongly stable semigroup, then on the imaginary axis A has neither eigenvalues nor a residual spectrum (see Arendt and Batty [2]). If, in addition, A has compact resolvent (so its spectrum consists only of eigenvalues), then the imaginary axis is contained in the resolvent set. Thus, in this case, the invertibility condition for A in the above theorem is satisfied.

Of course, our result is somewhat different from the one in Logemann, Ryan and Townley [70]. They did not assume that the system was strictly positive real, which forced them to use low gain control: the possible gains for the integral controller are related to those $k > 0$ for which $1 + k\mathrm{Re}[G(s)/s]$ is positive real. Our assumption that $G(s)$ is strictly positive real removes the necessity of using low gain control.

Chapter 9

Examples of Strongly Stabilizable Systems

In this chapter, we present four examples of models of simple engineering systems. All our examples are described by a system of partial differential equations and ordinary differential equations, and three of them have a state-space representation $\Sigma(A, B, B^*, D)$ on a Hilbert space Z, satisfying assumptions A1–A5 in Section 2.2. These examples show that models of partial differential equations from various physical domains and with various spatial dimensions can be cast into the $\Sigma(A, B, B^*)$ framework. The fourth example is a hybrid system consisting of the parallel connection of a $\Sigma(A, B, B^*)$ system with a finite-dimensional system.

The first example, which was introduced in Example 2.2.4, describes the transverse vibrations of a flexible beam, where control takes place through a piezoelectric film. The second example is a partial differential equation describing the vibrations of a two-dimensional object; it is a model of boundary control of a vibrating rectangular plate. The third example describes the control of sound in a one-dimensional wave-guide, where the ends react as linear oscillators to the acoustic pressure. We have worked out the approximation procedure for solutions of the standard LQ Riccati equation described in Chapter 5 for this example. The fourth and last example is a general model for a flexible structure with rigid body dynamics.

Note that even though the first three models have boundary control and boundary observation, they are still described by state-space models with a bounded control operator B and a bounded observation operator C (which will turn out to be equal to B^*).

9.1 Boundary Control of a Flexible Beam

This section studies Example 2.2.4 in more detail. The model of this example describes the transverse vibrations of a beam of length L, which is clamped at one end and has a point mass attached at the tip. A piezoelectric film is bonded to the beam, which applies a bending moment to the beam if a voltage is applied to it. This voltage is the control input of the system and the angular velocity at the tip is measured as output.

This system was described by Bailey and Hubbard [3] as a model of one of the arms of a satellite, consisting of a central hub with four flexible beams attached to it. In their paper, three different controllers were designed as active dampers for this system and tested in numerical simulations and on an experimental test structure. The model they used is given by the partial differential equation

$$\frac{\partial^4 w}{\partial x^4}(x,t) + \frac{\partial^2 w}{\partial t^2}(x,t) = 0 \text{ for } 0 < x < L, \tag{9.1}$$

with boundary conditions

$$w(0,t) = \frac{\partial w}{\partial x}(0,t) = 0, \tag{9.2}$$

$$\frac{\partial^2 w}{\partial x^2}(L,t) = -\frac{\partial^3 w}{\partial t^2 \partial x}(L,t) + u(t), \tag{9.3}$$

$$\frac{\partial^3 w}{\partial x^3}(L,t) = \frac{\partial^2 w}{\partial t^2}(L,t) \tag{9.4}$$

and measurement

$$y(t) = \frac{\partial^2 w}{\partial t \partial x}(L,t). \tag{9.5}$$

In Slemrod [94], a state-space representation of this model was provided and analyzed. To obtain such a first-order representation, a number of spaces is needed. The input space and output space both equal $\mathbb{R}$ and are denoted by U. Let $z = \text{col}(z_1, z_2, z_3, z_4)$. The state space is the Hilbert space

$$Z = \left\{ z \in H^2(0,L) \times \mathbf{L}_2(0,L) \times \mathbb{R} \times \mathbb{R} \mid z_1 = \frac{\partial z_1}{\partial x} = 0 \text{ at } x = 0 \right\}$$

with the inner product

$$\langle z, \tilde{z} \rangle_Z = \int_0^L \frac{\partial^2 z_1}{\partial x^2}(x) \frac{\partial^2 \tilde{z}_1}{\partial x^2}(x) dx + \int_0^L z_2 \tilde{z}_2 dx + z_3 \tilde{z}_3 + z_4 \tilde{z}_4.$$

We define the operators $A : \mathcal{D}(A) \subset Z \to Z$ and $B \in \mathcal{L}(U,Z)$ by

$$\mathcal{D}(A) = \left\{ z \in H^4(0,L) \times H^2(0,L) \times \mathbb{R} \times \mathbb{R} \mid \right.$$

$$\left. z_1(0) = \frac{\partial z_1}{\partial x}(0) = 0, \ z_2(0) = \frac{\partial z_2}{\partial x}(0) = 0, \ z_2(L) = z_3, \ \frac{\partial z_2}{\partial x}(L) = z_4 \right\},$$

$$A = \begin{bmatrix} 0 & I & 0 & 0 \\ -\dfrac{\partial^4}{\partial x^4} & 0 & 0 & 0 \\ \dfrac{\partial^3}{\partial x^3}\big|_{x=L} & 0 & 0 & 0 \\ -\dfrac{\partial^2}{\partial x^2}\big|_{x=L} & 0 & 0 & 0 \end{bmatrix}, \qquad B = \begin{bmatrix} 0 \\ 0 \\ 0 \\ 1 \end{bmatrix}.$$

With these definitions of the Hilbert space Z and the operators A and B, the system (9.1)–(9.5) can be represented by the abstract differential equation on Z,

$$\begin{aligned} \dot{z}(t) &= Az(t) + Bu(t), \quad z(0) = z_0, \\ y(t) &= B^* z(t). \end{aligned} \tag{9.6}$$

Slemrod [94] has shown that A is a closed, densely defined, skew-adjoint operator with compact resolvent (in fact, A is invertible and A^{-1} is compact). Furthermore, A generates a C_0-semigroup of contractions $T(t)$ on Z. Slemrod computed the eigenvalues of A and showed that there exists an orthonormal basis in Z of eigenvectors of A. The eigenvalues of A are purely imaginary and of the form $\lambda = \pm j\mu^2$, where μ satisfies the equation

$$0 = 1 + \cos(\mu)\cosh(\mu) + \mu[\sinh(\mu)\cos(\mu) - \cosh(\mu)\sin(\mu)]$$
$$- \mu^3[\cos(\mu)\sinh(\mu) + \sin(\mu)\cosh(\mu)] + \mu^4[1 - \cos(\mu)\cosh(\mu)].$$

Hence, it can be concluded that A is a Riesz-spectral operator (for details on Riesz-spectral operators, see Section 2.3 in Curtain and Zwart [36]). Slemrod showed that $\Sigma(A, -, B^*)$ is approximately observable for appropriate (generic) choices of the length of the beam L. Because A is a Riesz-spectral operator, we can use Theorem 4.2.3 in Curtain and Zwart to show that $\Sigma(A, -, B^*)$ is approximately observable if and only if $\Sigma(A^*, B, -)$ is approximately controllable. In our case, $A^* = -A$ and $\Sigma(A, -, B^*)$ is approximately observable, so we have that $\Sigma(-A, B, -)$ is approximately controllable. It is an easy application of Theorem 4.2.3 in Curtain and Zwart that $\Sigma(A, B, -)$ is then approximately controllable, for generic choices of L, as well.

Summarizing, the model (9.1)–(9.5) can be represented by $\Sigma(A, B, B^*)$ as in (9.6). Furthermore, $\Sigma(A, B, B^*)$ satisfies conditions A1–A5 in Section 2.2 and, as a consequence, it satisfies properties P1–P5. In particular, this means that this system is strongly stabilizable while it cannot be exponentially stabilizable, because B is of finite rank and bounded and A has infinitely many poles on the imaginary axis!

9.2 Boundary Control of a Flexible Plate

In this section, we look at the example of a partial differential equation on a two-dimensional spatial domain. The model is taken from You [120]. It describes the transverse vibrations of an elastic square plate with an edging mass along one edge. The model consists of the two-dimensional Petrovsky equation for the dynamics of the plate, coupled with the ordinary differential equations describing the dynamics at the edge where the mass is attached.

Consider a thin, homogeneous, square elastic plate with one clamped edge Γ_0 and two simply supported edges Γ_1 and Γ_3. The last edge, Γ_2, is subject to dynamical equations involving the inertial force, inertial torque, external control force $f_1(y, t)$ and external control torque $f_2(y, t)$. The latter two constitute the input $u(t) = \text{col}(f_1(t), f_2(t))$.

Let $\Omega = [0, 1] \times [0, 1]$ and let $w(x, y, t)$ be the transverse displacement of the plate for $(x, y) \in \Omega$. Define $\Gamma_0 = \{(x, y) \in \Omega : x = 0\}$, $\Gamma_1 = \{(x, y) \in \Omega : y = 0\}$, $\Gamma_2 = \{(x, y) \in \Omega : x = 1\}$ and $\Gamma_3 = \{(x, y) \in \Omega : y = 1\}$.

The following is a suitable model for the vibrating plate:

$$\frac{\partial^2 w}{\partial t^2}(x, y, t) + \Delta^2 w(x, y, t) = 0 \quad \text{for } (x, y) \in \Omega, \tag{9.7}$$

with initial conditions

$$\begin{aligned} w(x, y, 0) &= w_0, \\ \frac{\partial w}{\partial t}(x, y, 0) &= w_1 \quad \text{for } (x, y) \in \Omega \end{aligned} \tag{9.8}$$

and boundary conditions

$$w(0,y,t) = \frac{\partial w}{\partial x}(0,y,t) = 0,$$

$$w(x,0,t) = \frac{\partial^2 w}{\partial y^2}(x,0,t) = 0,$$

$$w(x,1,t) = \frac{\partial^2 w}{\partial y^2}(x,1,t) = 0, \tag{9.9}$$

$$\frac{\partial^2 w}{\partial t^2}(1,y,t) = \frac{\partial^3 w}{\partial x^3}(1,y,t) + (2-\sigma)\frac{\partial^3 w}{\partial x\partial y^2}(1,y,t) + f_1(y,t),$$

$$\frac{\partial^3 w}{\partial t^2\partial x}(1,y,t) = -\frac{\partial^2 w}{\partial x^2}(1,y,t) - \sigma\frac{\partial^2 w}{\partial y^2}(1,y,t) + f_2(y,t),$$

where σ is the Poisson ratio. If we measure the linear and angular velocities along Γ_2, we obtain the measurement

$$y(t) = \begin{bmatrix} \dfrac{\partial w}{\partial t}(1,y,t) \\[2mm] \dfrac{\partial^2 w}{\partial t\partial x}(1,y,t) \end{bmatrix}. \tag{9.10}$$

To obtain a suitable first-order formulation, we need to introduce a number of spaces and operators. The following Hilbert space will be used to construct the state space:

$$Z_0 = \mathbf{L}_2(\Omega) \times \mathbf{L}_2(0,1) \times \mathbf{L}_2(0,1)$$

with inner product

$$\langle z,\tilde{z}\rangle_{Z_0} = \int_\Omega w(x,y)\tilde{w}(x,y)dxdy + \int_0^1 w_1(y)\tilde{w}_1(y)dy + \int_0^1 w_2(y)\tilde{w}_2(y)dy,$$

where $z = \text{col}(w(x,y),w_1(y),w_2(y)) \in Z_0$. We define $A_0 : \mathcal{D}(A_0) \to Z_0$ by

$$\mathcal{D}(A_0) = \Big\{ z = \text{col}(w(x,y),w_1(y),w_2(y)) \in H^4(\Omega) \times \mathbf{L}_2(0,1) \times \mathbf{L}_2(0,1)|$$

$$w(0,y) = \frac{\partial w}{\partial x}(0,y) = 0,\ w(1,y) = w_1(y),\ \frac{\partial w}{\partial x}(1,y) = w_2(y),$$

$$w(x,0) = \frac{\partial^2 w}{\partial y^2}(x,0) = w(x,1) = \frac{\partial^2 w}{\partial y^2}(x,1) = 0 \Big\},$$

$$A_0 = \begin{bmatrix} \Delta^2 & 0 & 0 \\[2mm] -\left(\dfrac{\partial^3}{\partial x^3} + (2-\sigma)\dfrac{\partial^3}{\partial x\partial y^2}\right)\Big|_{x=1} & 0 & 0 \\[3mm] \left(\dfrac{\partial^2}{\partial x^2} + \sigma\dfrac{\partial^2}{\partial y^2}\right)\Big|_{x=1} & 0 & 0 \end{bmatrix}.$$

Next, we introduce the Hilbert space Z_1 with inner product $\langle\ ,\ \rangle_{Z_1}$;

$$Z_1 = \mathcal{D}(A_0^{\frac{1}{2}}),$$

$$\langle z,\tilde{z}\rangle_{Z_1} = \left\langle A_0^{\frac{1}{2}}z, A_0^{\frac{1}{2}}\tilde{z}\right\rangle_{Z_0},$$

and the operator $A_1 : \mathcal{D}(A_1) \subset Z_1 \to Z_0$,

$$\mathcal{D}(A_1) = \{ z \in Z_1 : \exists \tilde{z} \in Z_0 \text{ such that } \langle z, h \rangle_{Z_1} = \langle \tilde{z}, h \rangle_{Z_0} \forall h \in Z_1 \},$$
$$A_1 z = \tilde{z}.$$

Finally, the state space is the Hilbert space $Z = Z_1 \times Z_0$ and the system operator $A : \mathcal{D}(A) \to Z$ is defined by

$$A = \begin{bmatrix} 0 & I \\ -A_1 & 0 \end{bmatrix}, \tag{9.11}$$

where $\mathcal{D}(A) = \mathcal{D}(A_1) \times Z_1$. The input space is the Hilbert space

$$U = \mathbf{L}_2(0,1) \times \mathbf{L}_2(0,1) \tag{9.12}$$

with the usual inner product, and the input operator $B : U \to Z$ is defined by

$$B = \begin{bmatrix} 0 \\ B_0 \end{bmatrix}, \tag{9.13}$$

where $B_0 : U \to Z_0$ is given by

$$B_0 = \begin{bmatrix} 0 & 0 \\ I & 0 \\ 0 & I \end{bmatrix}.$$

In You [120], it is shown that $A : \mathcal{D}(A) \subset Z \to Z$ is a closed, densely defined skew-adjoint operator with compact resolvent ($A^{-1} \in \mathcal{L}(Z)$ is compact). A generates a unitary C_0-group of operators $T(t)$ on Z. The system (9.7)–(9.10) can be reformulated as

$$\dot{z}(t) = Az(t) + Bu(t),$$
$$y(t) = B^* z(t)$$

for $z(t) \in Z$, with u and $y \in \mathbf{L}_2(0,1) \times \mathbf{L}_2(0,1)$ and $B \in \mathcal{L}(U, Z)$.

In the same paper, it is proved that $A - BB^*$ generates a strongly stable contraction semigroup $T_B(t)$. Since A has compact resolvent, this implies that $T_B^*(t)$ is also strongly stable. You also shows that $\Sigma(A, B, -)$ is approximately controllable by appealing to results in Benchimol [19]. Here, we prove approximate controllability and observability by appealing to properties of Lyapunov equations from Hansen and Weiss [53] (see also our Lemma 2.1.4). It is clear that the following Lyapunov equations have the solution $\frac{1}{2} I$:

$$(A - BB^*)^* L_C z + L_C (A - BB^*) z = -BB^* z, \quad z \in \mathcal{D}(A), \tag{9.14}$$
$$(A - BB^*) L_B z + L_B (A - BB^*)^* z = -BB^* z, \quad z \in \mathcal{D}(A^*). \tag{9.15}$$

Since $T_B^*(t)$ and $T_B(t)$ are strongly stable, $L_C = \frac{1}{2} I = L_B$ are the unique solutions of (9.14) and (9.15), respectively. From Lemma 2.1.4, we conclude that $\Sigma(A - BB^*, B, B^*)$ have identical controllability and observability Gramians $L_C = \frac{1}{2} I = L_B$. As in Theorem 4.1.22 of Curtain and Zwart [36], this shows that $\Sigma(A - BB^*, B, B^*)$ is both approximately controllable and observable. Since these properties are invariant under feedback and output injection, respectively, this implies that $\Sigma(A, B, B^*)$ is both approximately controllable and observable. The above analysis shows that this state-space realization satisfies assumptions A1–A5 in Section 2.2.

9.3   Propagation of Sound in a Duct

Our third example describes an acoustical system. It concerns the (one-dimensional) propagation of sound in a duct. The ends of the duct act as a linear oscillator to the acoustic pressure inside. Actuating and sensing takes place through these linear oscillators. A system of this kind can be used to model frequency-dependent reflection of an acoustic plane wave at a so-called locally reacting surface (see page 263 of Morse and Ingard [78]). A number of similar models with different acoustical boundary conditions is described in Banks, Propst and Silcox [12].

We assume that the flow is uniform in a cross section of the duct. Hence, the sound propagates only in the longitudinal direction and the quantities depend only on the longitudinal spatial variable x. $\Psi(x,t)$ denotes the velocity potential, so that $\partial\Psi/\partial x(x,t)$ is the particle velocity in the fluid and $\partial\Psi/\partial t(x,t)$ is the acoustic pressure. Now, Ψ satisfies the wave equation

$$\frac{\partial^2\Psi}{\partial t^2}(x,t) = \frac{\partial^2\Psi}{\partial x^2}(x,t), \ \ x \in (0,1).$$

The displacements of the two ends $\eta_0(t)$ at $x=0$ and $\eta_1(t)$ at $x=1$ satisfy

$$\frac{d^2\eta_0}{dt^2}(t) + d_0\frac{d\eta_0}{dt}(t) + k_0\eta_0(t) = -\rho_0\frac{\partial\Psi}{\partial t}(0,t) + f_1(t),$$

$$\frac{d^2\eta_1}{dt^2}(t) + d_1\frac{d\eta_1}{dt}(t) + k_1\eta_1(t) = -\rho_1\frac{\partial\Psi}{\partial t}(1,t) + f_2(t),$$

where f_1 and f_2 are external forces at the ends. From continuity of velocity at the boundary, we obtain

$$\frac{d\eta_0}{dt}(t) = -\frac{\partial\Psi}{\partial x}(0,t),$$

$$\frac{d\eta_1}{dt}(t) = \frac{\partial\Psi}{\partial x}(1,t).$$

The measurements are taken to be proportional to the velocities at the endpoints,

$$y_1(t) = \frac{1}{2\rho_0}\frac{d\eta_0}{dt}(t),$$

$$y_2(t) = \frac{1}{2\rho_1}\frac{d\eta_1}{dt}(t).$$

The model could be brought into first-order form straightforwardly by introducing $z = \mathrm{col}(z_1,z_2,z_3,z_4)$ as state vector, where $z_1 = \Psi$, $z_2 = \partial\Psi/\partial t$, $z_3 = \mathrm{col}(\eta_0,\eta_1)$ and $z_4 = \mathrm{col}(\dot\eta_0,\dot\eta_1)$. The state space is then endowed with the norm corresponding to the energy of the system (see Beale [17]). For the approximation that we want to perform, a different state-space realization introduced by Ito and Propst [56] is more convenient. An important difference between the approach of Ito and Propst and that of Beale is that the former authors obtain an A-operator that does have compact resolvent, but Beale does not. The approach of Ito and Propst is to decompose Ψ into a part w^+ propagating in the positive direction and a part w^- propagating in the negative direction:

$$w^+(x,t) = \frac{1}{2}\left(\frac{\partial\Psi}{\partial t}(x,t) - \frac{\partial\Psi}{\partial x}(x,t)\right),$$

$$w^-(x,t) = \frac{1}{2}\left(\frac{\partial\Psi}{\partial t}(x,t) + \frac{\partial\Psi}{\partial x}(x,t)]\right).$$

Defining $\phi_0 = \eta_0$, $\phi_1 = \eta_1$, $\psi_0 = \dot{\eta}_0$, $\psi_1 = \dot{\eta}_1$, we obtain

$$
\frac{d}{dt}
\begin{bmatrix}
w^-(x,t) \\
w^+(x,t) \\
\phi_0(t) \\
\phi_1(t) \\
\psi_0(t) \\
\psi_1(t)
\end{bmatrix}
=
\begin{bmatrix}
\dfrac{\partial w^-}{\partial x}(x,t) \\[2mm]
-\dfrac{\partial w^+}{\partial x}(x,t) \\[2mm]
\psi_0(t) \\
\psi_1(t) \\
-\rho_0 w^-(0,t) - \rho_0 w^+(0,t) - k_0 \phi_0(t) - d_0 \psi_0(t) \\
-\rho_1 w^-(1,t) - \rho_1 w^+(1,t) - k_1 \phi_1(t) - d_1 \psi_1(t)
\end{bmatrix},
$$

with boundary conditions

$$
\psi_0(t) = -w^-(0,t) + w^+(0,t),
$$
$$
\psi_1(t) = w^-(1,t) - w^+(1,t).
$$

The state space is taken to be $Z_0 = \mathbf{L}_2(0,1) \times \mathbf{L}_2(0,1) \times \mathbb{R}^4$ with inner product (writing $z = \text{col}(w^-(x), w^+(x), \phi_0, \phi_1, \psi_0, \psi_1)$)

$$
\langle z, \tilde{z} \rangle = \langle w^-, \tilde{w}^- \rangle_{L_2} + \langle w^+, \tilde{w}^+ \rangle_{L_2} + \frac{k_0}{2\rho_0} \phi_0 \tilde{\phi}_0 + \frac{k_1}{2\rho_1} \phi_1 \tilde{\phi}_1 \qquad (9.16)
$$
$$
+ \frac{1}{2\rho_0} \psi_0 \tilde{\psi}_0 + \frac{1}{2\rho_1} \psi_1 \tilde{\psi}_1.
$$

The input and output spaces are $U = Y = \mathbb{R}^2$. Let us define the operators A_0, B_0 and C_0 as

$$
\mathcal{D}(A_0) = \{ z \in Z_0 \mid w^- \in H^1(0,1),\ w^+ \in H^1(0,1), \qquad (9.17)
$$
$$
\psi_0 = -w^-(0) + w^+(0),\ \psi_1 = w^-(1) - w^+(1) \},
$$

$$
A_0 z =
\begin{bmatrix}
\dfrac{\partial w^-}{\partial x}(x,t) \\[2mm]
-\dfrac{\partial w^+}{\partial x}(x,t) \\[2mm]
\psi_0(t) \\
\psi_1(t) \\
-\rho_0 w^-(0,t) - \rho_0 w^+(0,t) - k_0 \phi_0(t) - d_0 \psi_0(t) \\
-\rho_1 w^-(1,t) - \rho_1 w^+(1,t) - k_1 \phi_1(t) - d_1 \psi_1(t)
\end{bmatrix},
\qquad (9.18)
$$

$$
B_0 =
\begin{bmatrix}
0 & 0 \\
0 & 0 \\
0 & 0 \\
0 & 0 \\
1 & 0 \\
0 & 1
\end{bmatrix},
\quad
C_0 =
\begin{bmatrix}
0 & 0 & 0 & 0 & \dfrac{1}{2\rho_0} & 0 \\[2mm]
0 & 0 & 0 & 0 & 0 & \dfrac{1}{2\rho_1}
\end{bmatrix}.
\qquad (9.19)
$$

With these definitions, the system can be written as

$$\dot{z} = A_0 z + B_0 u,$$
$$y = C_0 z.$$

It is an easy computation to see that $C_0 = B_0^*$.

In Ito and Propst [56], it was shown that A_0 generates a C_0-semigroup of contractions $T_0(t)$ on Z_0 and A_0 has compact resolvent. Because A_0 has compact resolvent, A_0 has only point spectrum and it can be computed that $\lambda \in \sigma(A_0)$ if and only if λ satisfies

$$\begin{aligned}
0 = {} & (\lambda^2 + (d_0 + \rho_0)\lambda + k_0)(\lambda^2 + (d_1 + \rho_1)\lambda + k_1)e^{\lambda} \\
& - (\lambda^2 + (d_0 - \rho_0)\lambda + k_0)(\lambda^2 + (d_1 - \rho_1)\lambda + k_1)e^{-\lambda}.
\end{aligned} \tag{9.20}$$

All the eigenvalues have negative real part and occur in complex conjugate pairs. The associated eigenvectors are given by

$$z_\lambda = \begin{bmatrix}
(\lambda^2 + (d_0 + \rho_0)\lambda + k_0)(\lambda^2 + (d_1 - \rho_1)\lambda + k_1)e^{\lambda x} \\
(\lambda^2 + (d_0 - \rho_0)\lambda + k_0)(\lambda^2 + (d_1 - \rho_1)\lambda + k_1)e^{-\lambda x} \\
-2\rho_0(\lambda^2 + (d_1 - \rho_1)\lambda + k_1) \\
-2\rho_1(\lambda^2 + (d_0 + \rho_0)\lambda + k_0)e^{\lambda} \\
-2\rho_0(\lambda^2 + (d_1 - \rho_1)\lambda + k_1)\lambda \\
-2\rho_1(\lambda^2 + (d_0 + \rho_0)\lambda + k_0)\lambda e^{\lambda}
\end{bmatrix}.$$

The vectors $\{z_\lambda \mid \lambda \in \sigma(A_0)\}$ form a complete orthogonal set in Z_0, and all eigenvalues are simple.

Because, in acoustics, one is interested only in variations of the pressure, we have to add an extra condition to "filter out" states that correspond to the hydrostatic pressure, i.e., the part of pressure that is constant in the spatial variable x. It can be checked that this corresponds to states that are in the subspace spanned by the eigenvector corresponding to $\lambda = 0$, i.e., multiples of $z_0 = \text{col}(k_0 k_1, k_0 k_1, -2\rho_0 k_1, -2\rho_1 k_0, 0, 0)$. This leads to the extra condition

$$\int_0^1 w^-(x)dx + \int_0^1 w^+(x)dx - \phi_0 - \phi_1 = 0.$$

Therefore, we do not use Z_0 as our state space, but the quotient space

$$Z = \left\{ z \in Z_0 \,\middle|\, \int_0^1 w^-(x)dx + \int_0^1 w^+(x)dx - \phi_0 - \phi_1 = 0 \right\}.$$

Z is a closed, linear subspace of Z_0 and thus a Hilbert space with the same inner product. Define $A = A_0|_Z$. A is the generator of a C_0-semigroup of contractions $T(t)$ in Z because $T_0(t)$ maps Z into Z (this follows from the fact that z_0 is an eigenvector of A and Z is the orthogonal complement of z_0 in Z_0).

A also inherits the properties that it is a Riesz-spectral operator and that it has compact resolvent from A_0. So, A has only a point spectrum and, furthermore, $\lambda \in \sigma(A)$ if and only if $\lambda \neq 0$ and λ satisfies (9.20). We define B and C as the restrictions of B_0 and C_0 to Z. Using the controllability and observability test in Theorem 4.2.3 of Curtain and Zwart [36], it can be computed that $\Sigma(A, B, C)$ is approximately controllable and approximately observable.

We can summarize the previous results by saying that $C = B^*$ and the system $\Sigma(A, B, B^*)$ satisfies assumptions A1–A5 in Section 2.2.

For this example, we perform a numerical approximation of the strongly stabilizing solution of the standard LQ algebraic Riccati equation,

$$A^*Xz + XAz - XBB^*Xz + BB^*z = 0, \qquad z \in \mathcal{D}(A), \qquad (9.21)$$

as was described in Chapter 5. To do so, we construct a sequence of approximating finite-dimensional systems, and we will show that it satisfies the conditions that guarantee convergence of the approximating solutions of the finite-dimensional LQ Riccati equations to the solution of the infinite-dimensional LQ Riccati equation. Ito and Propst [56] proposed an approximation scheme for $\Sigma(A_0, B_0, C_0)$. We explain this scheme and then make an obvious modification, to obtain an approximation scheme for $\Sigma(A, B, C)$.

Let $z = \text{col}(w^-(x), w^+(x), \phi_0, \phi_1, \psi_0, \psi_1) \in Z_0$. We define z_0^N by

$$z_0^N = \pi_0^N z = (w_0^-, \dots, w_N^-, w_0^+, \dots, w_{N-1}^+, \phi_0^N, \phi_1^N, \psi_0^N, \psi_1^N)^T \in \mathbb{R}^{2N+5},$$

where $\phi_0^N = \phi_0, \phi_1^N = \phi_1, \psi_0^N = \psi_0, \psi_1^N = \psi_1$ and $w_k^-, w_m^+, k = 0, \dots, N, m = 0, \dots, N-1$, are defined as follows. Let $P_k(x)$ be the Legendre polynomial of degree k. The set $\{P_k(x), \ k \in \mathbb{N}\}$ forms an orthogonal basis of $\mathbf{L}_2(-1,1)$ and $\|P_k\|^2_{L_2(-1,1)} = 2/(2k+1)$. Hence, the polynomials $p_k(x) = \sqrt{2k+1}P_k(2x-1)$ form an orthonormal basis for $\mathbf{L}_2(0,1)$. We expand $w^-(x)$ and $w^+(x)$ with respect to the basis $\{p_k(x), \ k \in \mathbb{N}\}$ and truncate these expressions:

$$w_N^-(x) = \sum_{k=0}^{N} w_k^- p_k(x),$$

$$w_{N-1}^+(x) = \sum_{k=0}^{N-1} w_k^+ p_k(x).$$

Note the difference in notation between the approximating function $w_N^-(x)$ and the coefficient w_N^-. Let i_0^N denote the corresponding embedding that associates with a vector v in $\mathbb{R}^{2N+5}$ an element in Z_0 defined by

$$i_0^N v = \text{col}\left(\sum_{k=0}^{N} v_{k+1} p_k(x), \sum_{k=0}^{N-1} v_{N+2+k} p_k(x), v_{2N+2}, v_{2N+3}, v_{2N+4}, v_{2N+5} \right).$$

Next, we introduce

$$\tilde{w}_N^-(x) = \sum_{k=0}^{N} w_k^- p_k(x) + a p_{N+1}(x),$$

$$\tilde{w}_{N-1}^+(x) = \sum_{k=0}^{N-1} w_k^+ p_k(x) + b p_N(x),$$

where a and b are chosen such that $\tilde{w}_N^-(x)$ and $\tilde{w}_{N-1}^+(x)$ satisfy the boundary conditions

$$\psi_0 = -\tilde{w}_N^-(0) + \tilde{w}_{N-1}^+(0),$$

$$\psi_0 = \tilde{w}_N^-(1) - \tilde{w}_{N-1}^+(1).$$

We will now use $\tilde{w}_N^-(x)$ and $\tilde{w}_{N-1}^+(x)$ to define the approximation A_0^N of A_0:

$$
A_0^N
\begin{bmatrix}
w_0^- \\
\vdots \\
w_N^- \\
w_0^+ \\
\vdots \\
w_{N-1}^+ \\
\phi_0 \\
\phi_1 \\
\psi_0 \\
\psi_1
\end{bmatrix}
=
\begin{bmatrix}
a_0^- \\
\vdots \\
a_N^- \\
a_0^+ \\
\vdots \\
a_{N-1}^+ \\
\psi_0 \\
\psi_1 \\
-k_0\phi_0 - d_0\psi_0 - \rho_0\tilde{w}_N^-(0) - \rho_0\tilde{w}_{N-1}^+(0) \\
-k_1\phi_1 - d_1\psi_1 - \rho_1\tilde{w}_N^-(1) - \rho_1\tilde{w}_{N-1}^+(1)
\end{bmatrix},
$$

and $a_0^-, \ldots, a_N^-, a_0^+, \ldots, a_{N-1}^+$ are defined via

$$
\frac{d\tilde{w}_N^-}{dx}(x) = \sum_{k=0}^{N} a_k^- p_k(x), \qquad -\frac{d\tilde{w}_{N-1}^+}{dx}(x) = \sum_{k=0}^{N-1} a_k^+ p_k(x).
$$

The approximations B_0^N and C_0^N of B_0 and C_0 are given by

$$
B_0^N = \begin{bmatrix} 0_{(2N+3)\times 2} \\ I_2 \end{bmatrix}, \qquad C_0^N = (B^N)^*.
$$

The approximating state z^N is now obtained by orthogonal projection of z_0^N on the 2N+4-dimensional linear subspace of $\mathbb{R}^{2N+5}$ of vectors satisfying

$$
\int_0^1 \sum_{k=0}^{N} w_k^- p_k(x)dx + \int_0^1 \sum_{k=0}^{N-1} w_k^+ p_k(x)dx - \phi_0 - \phi_1 = 0.
$$

Let this orthogonal projection be denoted by χ^N and the associated embedding from $\mathbb{R}^{2N+4}$ into $\mathbb{R}^{2N+5}$ by h^N. The approximating system $\Sigma(A^N, B^N, C^N)$ is obtained by restricting A_0^N, B_0^N and C_0^N to this 2N+4-dimensional space, i.e., $A^N = \chi^N A_0^N h^N$, $B^N = \chi^N B_0^N$, $C^N = C_0^N h^N$. The projection π^N and embedding i^N for the total approximation are now given as

$$
i^N : \mathbb{R}^{2N+4} \to Z, \quad i^N = h^N i_0^N,
$$
$$
\pi^N : Z \to \mathbb{R}^{2N+4}, \quad \pi^N = \pi_0^N \chi^N.
$$

Now, $\pi^N i^N$ is an orthogonal projection in Z and $i^N \pi^N = I_{2N+4}$.

Note that the approximation A^N does not satisfy $A^N = \pi^N A i^N$, as in the scheme of Section 5. However, we do have $A^N + (A^N)^* \leq 0$ and $C^N = (B^N)^*$. So, the result of Lemma 5.4.1 is still valid.

In Ito and Propst [56], it was shown that for every $z \in \mathcal{D}(A_0)$ there exists a sequence $z^N \in \mathbb{R}^{2N+5}$ such that

$$
\|i^N z^N - z\| \to 0
$$

and

$$\|i^N A^N z^N - Az\| \to 0$$

as $N \to \infty$. This is easily adapted to the existence of a sequence $z^N \in \mathbb{R}^{2N+4}$ converging to $z \in \mathcal{D}(A)$. Similarly, we can prove that for all $z \in \mathcal{D}(A^*)$ there exists a sequence $z^N \in \mathbb{R}^{2N+4}$ such that

$$\|i^N z^N - z\| \to 0$$

and

$$\|i^N (A^N)^* z^N - A^* z\| \to 0.$$

It is easily shown that C^N and B^N converge strongly to C and B, respectively. As in Lemma 5.4.2, these results imply that the finite-dimensional systems $\Sigma(A^N, B^N, C^N)$ converge strongly to the infinite-dimensional system $\Sigma(A, B, B^*)$. Hence, we can conclude that if $\Sigma(A^N, B^N, C^N)$ is observable for all N, then the solution X of the infinite-dimensional LQ Riccati equation associated with $\Sigma(A, B, B^*)$ can be approximated by the sequence of solutions X^N of the matrix LQ Riccati equations associated with the systems $\Sigma(A^N, B^N, C^N)$.

We implemented the approximation scheme in Matlab. The values for the parameters we used are

$$\rho_0 = \rho_1 = 1,$$
$$d_0 = d_1 = 0.01,$$
$$k_0 = k_1 = 1.$$

We computed the approximating systems for $N = 2, 4, 8, 16, 32, 64$. Then, we computed the solutions to the Riccati equations corresponding to these approximations. To illustrate the convergence of the solutions, we compare the solutions in a number of characteristics. First, we compute the matrix norms of the approximating solutions. Second, we restrict X^{64} to the spaces on which the lower-order approximations are defined and compute the norm of the difference, $\|X^N - \pi^N i^{64} X^{64} \pi^{64} i^N\|$, for $N = 2, 4, 8, 16, 32$. This norm measures to what extent the lower order approximation matches the behavior of the higher order one on the lower-dimensional space. These values are listed in Table 9.1. Finally, we investigated the closed-loop systems corresponding to the different approximative Riccati feedbacks F_{X^N}. We have given plots of the closed-loop poles and we have computed the open-loop and closed-loop step responses. Because the step responses for the different values of N were indistinguishable, we give only the plots for $N = 16$.

N	$\|X^N\|$	$\|X^N - \pi^N i^M X^M \pi^M i^N\|$
2	0.9802	6.31e-11
4	0.9802	1.81e-10
8	0.9802	6.25e-10
16	0.9802	2.32e-9
32	0.9802	1.57e-8
64	0.9802	0

Table 9.1: Comparison of norms of the different approximations.

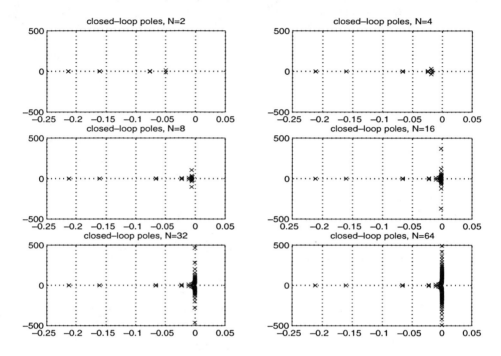

Figure 9.1: Closed-loop poles, $N = 2, 4, 8, 16, 32, 64$.

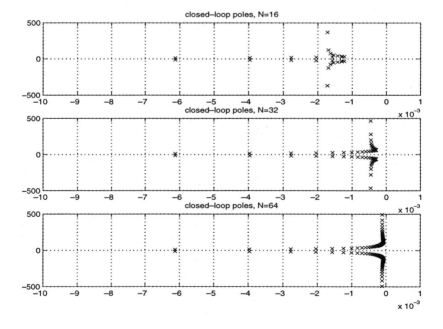

Figure 9.2: Close-up plot of closed-loop poles, $N = 16, 32, 64$.

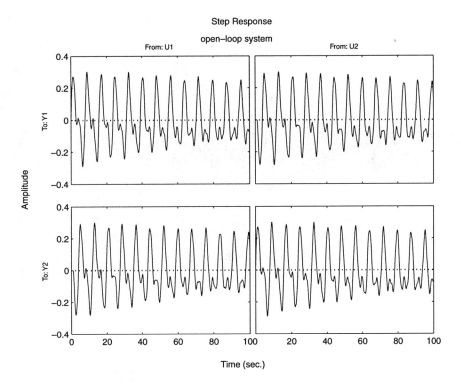

Figure 9.3: Open-loop step response, $N = 16$.

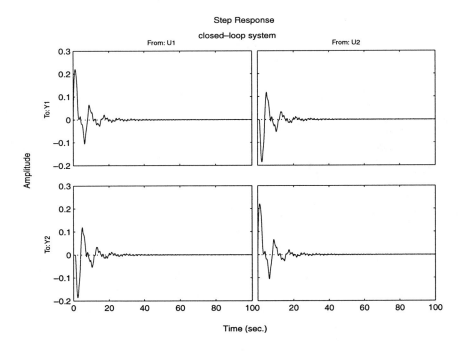

Figure 9.4: Closed-loop step response, $N = 16$.

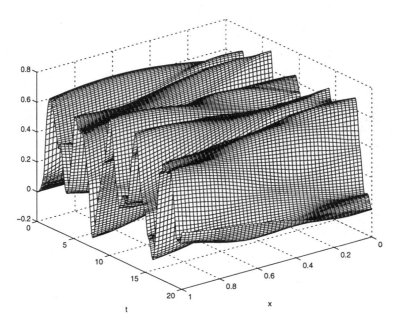

Figure 9.5: Open-loop step response, $N = 16$.

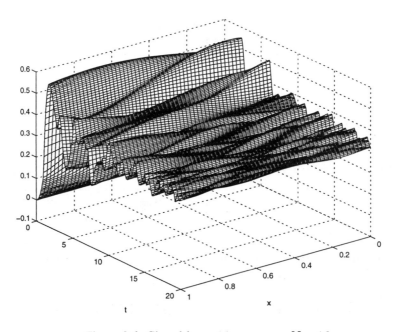

Figure 9.6: Closed-loop step response, $N = 16$.

From these results, we observe the following. The norms of the approximations are very close to each other and far larger than the norm of the difference between the low order approximations and the restriction of X^{64} to the lower order spaces. This indicates that the low order approximations succeed very well in capturing the behavior of the infinite-dimensional solutions on the finite-dimensional spaces on which they are defined. The plots of the closed-loop poles in Figures 9.1 and 9.2 indicate that the closed-loop system matrices converge as N tends to infinity. As expected, since the original system is not exponentially stabilizable, the poles are converging to the imaginary axis. The plots of the step responses for the different values of N are indistinguishable. We plotted step responses for $N = 16$ in Figures 9.3 and 9.4. In Figures 9.5 and 9.6, we plotted the response of the velocity potential $\Psi(x,t)$ to a step input at $x = 0$, for both open-loop and closed-loop systems.

9.4 A Flexible Structure with Rigid Body Dynamics

This last example gives a general model describing vibrations in a flexible structure combined with the rigid body dynamics of the structure (i.e., the translation of its center of mass and its rigid body rotation). The model we give is an infinite-dimensional version of the one in Sections 2.1 and 2.2 in Joshi [57].

This model is an example of a statically stabilizable system that is not in the $\Sigma(A, B, B^*)$ class for two reasons. First, A does not generate a contraction semigroup, and second, C is not equal to B^*. It is therefore a motivating example for those results in this book that deal with strong stability for systems $\Sigma(A, B, C)$ outside the $\Sigma(A, B, B^*)$ class.

The structure is equipped with a number (m_f) of force actuators and a number (m_T) of torque actuators. Let $\xi \in \mathbb{R}^3$ denote the position of the center of mass and let the force exerted by the ith actuator be $f_i \in \mathbb{R}^3$. The translation of the center of mass is given by

$$m\ddot{\xi} = \sum_{i=1}^{m_f} f_i. \tag{9.22}$$

The attitude of the structure is given by the vector $\alpha \in \mathbb{R}^3$ describing the rigid body Euler angle with respect to the x-, y- and z-axes. It satisfies

$$J\ddot{\alpha} = \sum_{i=1}^{m_f} r_i \times f_i + \sum_{j=1}^{m_T} T_j, \tag{9.23}$$

where T_j denotes the torque exerted by the jth torque actuator and $r_i \in \mathbb{R}^3$ describes the location where the force f_i is applied. The elastic motion is described by

$$\ddot{q} + D\dot{q} + \Lambda q = \sum_{i=1}^{m_f} \Psi_i^T f_i + \sum_{j=1}^{m_T} \Phi_j^T T_j \tag{9.24}$$

$$= \Psi^T F + \Phi^T T.$$

Here, q is the infinite sequence of modal amplitudes, which is assumed to be an element of a Hilbert space Q (for instance, $Q = l_2(\mathbb{N})$). $F = \mathrm{col}(f_1, \ldots, f_{m_f})$ and $T = \mathrm{col}(T_1, \ldots, T_{m_T})$. The infinite matrices $\Psi_i \in \mathcal{L}(Q, \mathbb{R}^3)$ and $\Phi_j \in \mathcal{L}(Q, \mathbb{R}^3)$ are called the *mode shape matrices* and *mode slope matrices*, respectively. Λ is the *stiffness operator* and it is assumed that

$\Lambda = \Lambda^* > 0$. The *damping operator* D satisfies $D = D^* \geq 0$. Both D and Λ are closed and densely defined. In general, both are unbounded operators.

We call this a hybrid system because it consist of a finite-dimensional part (the dynamics of ξ and α) coupled with a distributed part (the elastic motion). The coupling consists of the fact that both parts are affected by the same control inputs f_i and T_j. The aim of this section is to show that under certain conditions this system has a statically stabilizable state-space realization.

We define $p = \mathrm{col}(m\xi, J\alpha, q) \in \mathbb{R}^3 \times \mathbb{R}^3 \times Q =: Z_1$. Then the dynamics are described as

$$\ddot{p} + \tilde{D}\dot{p} + \tilde{\Lambda}p = \Gamma^T u, \tag{9.25}$$

where

$$\tilde{D} = \begin{bmatrix} 0 & 0 & 0 \\ 0 & 0 & 0 \\ 0 & 0 & D \end{bmatrix}, \quad \tilde{\Lambda} = \begin{bmatrix} 0 & 0 & 0 \\ 0 & 0 & 0 \\ 0 & 0 & \Lambda \end{bmatrix}, \tag{9.26}$$

$$\Gamma^T = \begin{bmatrix} I_3 & \cdots & I_3 & 0_3 & \cdots & 0_3 \\ R_1 & \cdots & R_{m_f} & I_3 & \cdots & I_3 \\ \Psi_1^T & \cdots & \Psi_{m_f}^T & \Phi_1^T & \cdots & \Phi_{m_T}^T \end{bmatrix},$$

$u = \mathrm{col}(f_1, \ldots, f_{m_f}, T_1, \ldots, T_{m_T}) \in \mathbb{R}^{3m_f + 3m_T} =: U$ and the matrices R_i, $i = 1 \ldots, m_f$, are given by

$$R_i = \begin{bmatrix} 0 & -r_z^i & r_x^i \\ -r_z^i & 0 & -r_y^i \\ -r_x^i & r_y^i & 0 \end{bmatrix}, \quad \text{where } r_i = \begin{pmatrix} r_x^i \\ r_y^i \\ r_z^i \end{pmatrix}.$$

If we have position sensors at the same locations as the force actuators and attitude sensors at the same locations as the torque actuators, then the observation is given by

$$y_p = \Gamma p. \tag{9.27}$$

On the other hand, the use of rate sensors measuring translational and rotational velocities at those locations leads to the observation

$$y_r = \Gamma \dot{p}. \tag{9.28}$$

Following Joshi [57], let us suppose that both types of measurements are available and the observation is $y = \mathrm{col}(y_p, y_r)$. To construct a first-order representation, we take as the state $z = \mathrm{col}(z, \dot{z})$ on the state space $Z = \{z = \mathrm{col}(z_1, z_2) \in Z_1 \times Z_1\}$ with inner product

$$\langle z, w \rangle_Z = \langle (\tilde{\Lambda} + \Gamma^T \Gamma) z_1, w_1 \rangle_{Z_1} + \langle z_2, w_2 \rangle_{Z_1}. \tag{9.29}$$

This leads to the following representation of the systems (9.22), (9.23), (9.24), (9.27), (9.28),

$$\dot{z}(t) = Az(t) + Bu(t), \quad z(0) = z_0,$$
$$y(t) = Cz(t), \tag{9.30}$$

where

$$A = \begin{bmatrix} 0 & I \\ -\tilde{\Lambda} & -\tilde{D} \end{bmatrix}, \quad B = \begin{bmatrix} 0 \\ \Gamma^T \end{bmatrix}, \quad C = \begin{bmatrix} C_p \\ C_r \end{bmatrix} = \begin{bmatrix} \Gamma & 0 \\ 0 & \Gamma \end{bmatrix}.$$

To show that the above equations represent a bounded linear system, there is some work to be done. We need to show that

1. $C \in \mathcal{L}(Z, U \times U)$;

2. equation (9.29) defines an inner product on Z. This is true only if $\tilde{\Lambda} + \Gamma^T\Gamma > 0$;

3. A is the generator of a C_0-semigroup on Z.

Let us examine these three issues:

1. The operator Γ is a bounded operator from Z_1 to U because the operators Φ_i, Ψ_j and R_i are bounded. Therefore, it is obvious that $C_r = [\, 0 \;\; \Gamma \,]$ is bounded. To establish the boundedness of C, we therefore need only to check that $C_p = [\, \Gamma \;\; 0 \,] \in \mathcal{L}(Z, U)$. We have

$$\begin{aligned}
\|z\|_Z^2 &= \langle (\tilde{\Lambda} + \Gamma^T\Gamma)z_1, z_1 \rangle_{Z_1} + \|z_2\|_{Z_1}^2 \\
&\geq \langle (\tilde{\Lambda} + \Gamma^T\Gamma)z_1, z_1 \rangle_{Z_1} \\
&\geq \|\Gamma z_1\|_U^2 \\
&= \|C_p z\|_U^2.
\end{aligned}$$

So, C is indeed a bounded output operator.

2. Next, we prove that $\tilde{\Lambda} + \Gamma^T\Gamma > 0$. From the relation between $\tilde{\Lambda}$ and Λ in (9.26) and the fact that Λ is assumed to be a positive, self-adjoint operator, it follows that $\tilde{\Lambda} + \Gamma^T\Gamma > 0$ if and only if

$$\left[\, I_6 \;\; 0 \,\right] \Gamma^T\Gamma \begin{bmatrix} I_6 \\ 0 \end{bmatrix} > 0.$$

So, defining $\tilde{\Gamma} = \Gamma \begin{bmatrix} I_6 \\ 0 \end{bmatrix}$, we require that $\ker(\tilde{\Gamma}) = \{0\}$, where

$$\tilde{\Gamma} = \begin{bmatrix} I_3 & R_1 \\ \vdots & \vdots \\ I_3 & R_{m_f} \\ 0 & I_3 \\ \vdots & \vdots \\ 0 & I_3 \end{bmatrix}.$$

Suppose that there exists an $x \in \mathbb{R}^6$ such that $\tilde{\Gamma}x = 0$. Then,

$$\begin{pmatrix} x_1 \\ x_2 \\ x_3 \end{pmatrix} + R_i \begin{pmatrix} x_4 \\ x_5 \\ x_6 \end{pmatrix} = 0$$

and

$$\begin{pmatrix} x_4 \\ x_5 \\ x_6 \end{pmatrix} = 0$$

for $i = 1, \ldots, m_f$. Consequently, $x = 0$ and, indeed, $\tilde{\Lambda} + \Gamma^T\Gamma > 0$.

3. Next, we show that A generates a C_0-semigroup on Z via a perturbation technique. More precisely, we show that $A_2 := A - [\Gamma^T \ 0][0 \ \Gamma^T]^T$ generates a C_0-semigroup Z. As A is a bounded perturbation of A_2, this implies that A generates a semigroup on Z, as well. We need the following assumptions:

(a) $\tilde{\Lambda} + \Gamma^T \Gamma$ is boundedly invertible;

(b) $\mathcal{D}(\tilde{\Lambda}) \subset \mathcal{D}(\tilde{D})$ (or, equivalently, $\mathcal{D}(\Lambda) \subset \mathcal{D}(D)$).

The domain of definition of A_2 is $\mathcal{D}(A_2) = \mathcal{D}(A) = \mathcal{D}(\tilde{\Lambda}) \times \mathcal{D}(\tilde{\Lambda}^{\frac{1}{2}})$. Clearly, A_2 is densely defined, because $\tilde{\Lambda}$ is. First, we show that A_2 is closed. It is easy to see that under assumptions (a) and (b), the operator V given by

$$V = \begin{bmatrix} -(\tilde{\Lambda} + \Gamma^T \Gamma)^{-1} \tilde{D} & -(\tilde{\Lambda} + \Gamma^T \Gamma)^{-1} \\ I & 0 \end{bmatrix}$$

satisfies

- $V \in \mathcal{L}(Z)$,

- $A_2 V = V A_2 = I$,

- $\text{Im}(V) = \mathcal{D}(A_2)$.

From Theorem A.3.46 in Curtain and Zwart [36], we can now conclude that A_2 is a closed operator. Next, we show that both A_2 and its adjoint are dissipative operators. For any $z \in \mathcal{D}(A_2)$,

$$\langle A_2 z, z \rangle_Z = \langle (\tilde{\Lambda} + \Gamma^T \Gamma) z_2, z_1 \rangle_{Z_1} + \langle -(\tilde{\Lambda} + \Gamma^T \Gamma) z_1 - \tilde{D} z_2, z_2 \rangle_{Z_1}$$
$$= \langle -\tilde{D} z_2, z_2 \rangle_{Z_1}$$
$$\leq 0.$$

Hence, A_2 is a dissipative operator. In exactly the same way, it can be shown that A_2^*, which is given by

$$A_2^* = \begin{bmatrix} 0 & -I \\ \tilde{\Lambda} + \Gamma^T \Gamma & -\tilde{D} \end{bmatrix},$$

is dissipative, as well. We can now apply the Lumer–Phillips Theorem (see Section 1.4 in Pazy [85]) to show that A_2 generates a C_0-semigroup of contractions. As a consequence,

$$A = A_2 + \begin{bmatrix} 0 & 0 \\ \Gamma^T \Gamma & 0 \end{bmatrix}$$

generates a C_0-semigroup, as well (see Theorem 3.2.1 in Curtain and Zwart [36] for a proof of the fact that $A + D$ generates a C_0-semigroup on Z if A generates a C_0-semigroup on Z and D is a bounded operator on Z).

We now state our main result concerning this example.

Lemma 9.4.1 *Let the system $\Sigma(A, B, C)$ be defined as in (9.30) on the state space Z and assume that the following conditions are satisfied:*

(i) $\Lambda = \Lambda^* > 0$, $D = D^* \geq 0$ are closed, densely defined operators with $\mathcal{D}(D) \supset \mathcal{D}(\Lambda)$;

(ii) A has compact resolvent;

(iii) $\Sigma(A,B,C)$ *is approximately controllable;*

(iv) $\Sigma(A_2,B,C_r)$ *is approximately observable, where* $A_2 = A - \begin{bmatrix} \Gamma^T & 0 \end{bmatrix} \begin{bmatrix} 0 \\ \Gamma \end{bmatrix}$;

(v) $\tilde{\Lambda} + \Gamma^T\Gamma$ *is boundedly invertible and* $(\tilde{\Lambda} + \Gamma^T\Gamma)^{-1}\tilde{D}$ *is bounded.*

Then the system $\Sigma(A,B,C)$ *is statically stabilizable. In particular, the static output feedback law* $u = -G_p y_p - G_r y_r$ *renders the closed-loop system strongly stable for any positive definite matrices* G_p *and* G_r.

Proof The conditions *(i)* and *(v)* ensure that $\Sigma(A,B,C)$ is a well-defined, bounded, linear system. The proof consists of the following steps.

1. We first restrict to the case that $G_p = I$ and show that the preliminary feedback $u = -y_p + v$ leads to the system $\Sigma(A_2,B,C)$. This system has almost all features of systems in the $\Sigma(A,B,B^*)$ class. The only difference is that there are more outputs, namely, y_p.

2. Then we show, using the $\Sigma(A,B,B^*)$ structure and additional information on the system, that the feedback $v = -G_r y_r$ leads to a strongly stable, closed-loop system $\Sigma(A_{cl},B,C)$, where $A_{cl} = A - BC_p - BG_r C_r$.

3. Finally, we show how to include the case of general positive definite matrices G_p.

Let us go through these steps. For ease of reference, we write

$$\Sigma_1 = \Sigma(A,B,C), \quad \Sigma_2 = \Sigma(A_2,B,C_r) = \Sigma(A_2,B,B^*).$$

1. Application of the static output feedback law $u = -y_p + v$ to the system Σ_1 leads to $\Sigma(A_2,B,C)$, i.e., the system Σ_2 with the additional output $y_p = C_p z$. We show that Σ_2 is in the $\Sigma(A,B,B^*)$ class. Earlier in this section, we have already shown that A_2 generates a contraction semigroup. As approximate controllability is invariant under feedback, Σ_2 is approximately controllable by assumption *(iii)*. Assumption *(iv)* states that Σ_2 is approximately observable. That A_2 has compact resolvent follows from the fact that A has compact resolvent, as follows. We have

$$(sI - A_2)^{-1} = (sI - A)^{-1} + (sI - A)^{-1}\begin{bmatrix} 0 & 0 \\ -\Gamma^T\Gamma & 0 \end{bmatrix}(sI - A_2)^{-1}.$$

The first term in the right-hand side is compact for all $s \in \rho(A)$. For $s \in \rho(A) \cap \rho(A_2)$, the second term is the product of a compact operator with a bounded operator; hence it is compact. As both A and A_2 are dissipative, the open right half-plane is contained in $\rho(A) \cap \rho(A_2)$, so this intersection is not empty. Consequently, the right-hand side is compact and so A_2 has compact resolvent. Thus, Σ_2 satisfies A1–A5 (see page 17) and so it is in the $\Sigma(A,B,B^*)$ class.

2. As $\Sigma_2 = \Sigma(A_2,B,B^*)$ belongs to our standard class, it is strongly stabilized by the feedback $v = -y_r$. That this also happens for $v = -G_r y_r$ can be seen by rewriting this case as the feedback $v = -y$ for the system $\Sigma(A,BS,SB^*)$, where S is the square root of G_r. This system is also in the $\Sigma(A,B,B^*)$ class. Thus, we have that $\Sigma_3 = \Sigma(A_2 - BG_r C_r, B, C_r) = \Sigma(A_{cl},B,C_r)$ is strongly stable. That is, we have

- $A_{cl} := A_2 - BG_r C_r = A - BG_r C_r - BC_p$ generates a strongly stable semigroup;

- input stability: $B^*(sI - A_{cl}^*)^{-1}z \in \mathbf{H}_2(U)$ for all $z \in Z$;

- output stability: $C_r(sI - A_{cl})^{-1}z \in \mathbf{H}_2(U)$ for all $z \in Z$;

- input-output stability: $C_r(sI - A_{cl})^{-1}B \in \mathbf{H}_\infty(\mathcal{L}(U))$.

However, we are interested in the strong stability of the overall closed-loop system $\Sigma_{cl} = \Sigma(A_{cl}, B, C)$. Hence, it remains to show that, in addition to the above, output stability and input-output stability hold for the output y_p, as well. The remaining analysis of part 2 of this proof is therefore dedicated to proving

- output stability with respect to y_p: $C_p(sI - A_{cl})^{-1}z \in \mathbf{H}_2(U)$ for all $z \in Z$;

- input-output stability with respect to y_p: $C_p(sI - A_{cl})^{-1}B \in \mathbf{H}_\infty(\mathcal{L}(U))$.

To do so, we take a closer look at $(sI - A_{cl})^{-1}$. For convenience of notation, let us write $\Lambda_{cl} = \tilde{\Lambda} + \Gamma^T\Gamma$ and $D_{cl} = \tilde{D} + \Gamma^T G_r \Gamma$. Then,

$$
\begin{aligned}
(sI - A_{cl})^{-1} &= \begin{bmatrix} sI & -I \\ \Lambda_{cl} & sI + D_{cl} \end{bmatrix}^{-1} \\
&= \begin{bmatrix} (s^2 I + sD_{cl} + \Lambda_{cl})^{-1}(sI + D_{cl}) & (s^2 I + sD_{cl} + \Lambda_{cl})^{-1} \\ -(s^2 I + sD_{cl} + \Lambda_{cl})^{-1}\Lambda_{cl} & s(s^2 I + sD_{cl} + \Lambda_{cl})^{-1} \end{bmatrix}. \quad (9.31)
\end{aligned}
$$

If A_{cl} is not exponentially stable, then $(sI - A_{cl})^{-1} \notin \mathbf{H}_\infty$. However, from the output stability of Σ_3, we can conclude that

$$
\Gamma(s^2 I + sD_{cl} + \Lambda_{cl})^{-1}\Lambda_{cl}z_1 \in \mathbf{H}_2(U) \quad \forall z_1 \in Z_1, \quad (9.32)
$$

$$
s\Gamma(s^2 I + sD_{cl} + \Lambda_{cl})^{-1}z_1 \in \mathbf{H}_2(U) \quad \forall z_1 \in Z_1 \quad (9.33)
$$

and from the input-output stability,

$$
s\Gamma(s^2 I + sD_{cl} + \Lambda_{cl})^{-1}\Gamma^T \in \mathbf{H}_\infty(\mathcal{L}(U)). \quad (9.34)
$$

We now first prove output stability of Σ_{cl}, i.e., we show that $C_p(sI - A_{cl})^{-1}z \in \mathbf{H}_2(U)$ for all $z \in Z$. From (9.31), it follows that

$$
\begin{aligned}
C_p(sI - A_{cl})^{-1}z &= s\Gamma(s^2 I + sD_{cl} + \Lambda_{cl})^{-1}z_1 \\
&\quad + \Gamma(s^2 I + sD_{cl} + \Lambda_{cl})^{-1}D_{cl}z_1 + \Gamma(s^2 I + sD_{cl} + \Lambda_{cl})^{-1}z_2.
\end{aligned}
$$

The first term on the right-hand side is in $\mathbf{H}_2$ for all z_1, by (9.33). For the second term, we obtain from the assumption that Λ_{cl}^{-1} and $\Lambda_{cl}^{-1}D_{cl}$ are bounded and from (9.32) that, for all $z_1 \in Z_1$,

$$
\Gamma(s^2 I + sD_{cl} + \Lambda_{cl})^{-1}D_{cl}z_1 = \Gamma(s^2 I + sD_{cl} + \Lambda_{cl})^{-1}\Lambda_{cl} \cdot \Lambda_{cl}^{-1}D_{cl}z_1 \in \mathbf{H}_2(U).
$$

In a similar way, it can be proved that the last term on the right-hand side is in $\mathbf{H}_2(U)$ for all $z_2 \in Z_1$, using (9.32) and the boundedness of Λ_{cl}^{-1}.

Next, we prove the input-output stability of Σ_{cl}, i.e., we show that the closed-loop transfer function from u to y_p, $C_p(sI - A_{cl})^{-1}B$, is in $\mathbf{H}_\infty(\mathcal{L}(U))$. We have

$$
C_p(sI - A_{cl})^{-1}B = \Gamma(s^2 I + sD_{cl} + \Lambda_{cl})^{-1}\Gamma^T
$$

and we denote the above transfer function by $G(s)$. If we write

$$
F(s) = s\Gamma(s^2 I + sD_{cl} + \Lambda_{cl})^{-1}\Gamma^T,
$$

then $G(s) = F(s)/s$. By (9.34), $F(s) \in \mathbf{H}_\infty(\mathcal{L}(U))$. The function $1/s$ is also holomorphic in $\mathbb{C}_0^+$ and it is bounded in $\mathbb{C}_0^+ \backslash B_\varepsilon(0)$ for any $\varepsilon > 0$, where we define $B_\varepsilon(0) = \{s \in \mathbb{C}_0^+ \mid |s| < \varepsilon\}$. Combining these observations, it follows that $G(s)$ is holomorphic in $\mathbb{C}_0^+$ and bounded in $\mathbb{C}_0^+ \backslash B_\varepsilon(0)$. Furthermore, $G(0) = \Gamma \Lambda_{cl}^{-1} \Gamma^T$, which is bounded by assumption *(v)*. Thus, G does not have a pole at $s = 0$ and we can conclude that G is also bounded on $B_\varepsilon(0)$, which shows that $G \in \mathbf{H}_\infty(\mathcal{L}(U))$.

3. The case of general G_p can be dealt with by defining the inner product on the state space alternatively as

$$\langle z, w \rangle_{Z,new} = \langle (\tilde{\Lambda} + \Gamma^T G_p \Gamma) z_1, w_1 \rangle_{Z_1} + \langle z_2, w_2 \rangle_{Z_1}.$$

The whole analysis above goes through without change in this new inner product. From the positive definiteness of G_p, it can easily be shown that the norm induced by this new inner product is equivalent to the original norm. Hence, the system is also strongly stable with respect to the original state space Z □

Assumptions *(i)* and *(ii)* are satisfied by most, if not all, systems of interest. Assumption *(iii)* and *(iv)* depend on the actuators and sensors implemented. Note that assumption *(iv)* concerns A_2, not A. The reason is that the pair (A, C_r) can never be observable: the position of the center of mass and the attitude of the structure cannot be reconstructed from the measurements of the rate sensor, because the dynamics do not depend on the position and attitude. However, the coupling introduced in A_2 makes the dynamics dependent on them, and so it becomes possible to observe the position and the attitude from the rate sensors.

Although for certain structures of practical interest assumption *(v)* will be satisfied, it still is quite restrictive. We used it in the general setup of the example in this section because it leads to a fairly simple proof. We believe that in more specific examples this assumption can be relaxed.

Chapter 10

Conclusions

In the preceding chapters, we have presented a diversity of system and control theoretic results about strongly stabilizable control systems. Here, we give an overview of the main contributions of the book and we discuss some open problems and different research directions.

Strong stability: We have introduced a framework for studying strongly stabilizable control systems. This includes definitions of strong stability, strong stabilizability and strong detectability. An important feature of all those definitions is that they comprise both internal and external stability properties. Directly related to this framework are basic system theoretic results such as the equivalence of internal and external stability of strongly stabilizable and detectable systems and the convergence of the state of a strongly stable system when an $\mathbf{L}_2$-input is applied. Within this general framework, we have introduced a special class of systems, the class $\Sigma(A, B, B^*)$ of dissipative systems with collocated actuators and sensors. These systems are often used as models of flexible structures and other mechanical systems, and their special structure allows us to give stronger results for systems in this class.

Riccati equations: We have given conditions for the existence and uniqueness of a strongly stabilizing solution to very general algebraic Riccati equations. We have shown that a Riccati equation associated with a strongly stable system has a strongly stabilizing solution if and only if the Popov function has a Wiener–Hopf factorization. This result was adapted to include statically stabilizable systems. This factorization-based approach allowed us to give versions of the positive real and bounded real lemmas for strongly stable systems and to prove a result about the existence of J-spectral factorizations. For the LQ Riccati equation, we showed that there exists a unique, strongly stabilizing solution if the system is strongly stabilizable and strongly detectable. Furthermore, we developed a procedure for approximating the strongly stabilizing solution of the LQ Riccati equation by solutions of a sequence of finite-dimensional Riccati equations and we proved the convergence of this approximation procedure.

Coprime factorizations and dynamic compensators: We studied input-output stabilization and strong stabilization by dynamic compensators for strongly stabilizable and detectable systems. In connection with this, we studied the key question of state-space formulas for normalized doubly coprime factorizations for strongly stabilizable and detectable systems. It is doubtful that a satisfactory generalization of all the known finite-dimensional results can be obtained for strongly stabilizable and detectable systems. However, under the assumptions of compact resolvent and finite-dimensional input and output spaces, we have obtained state-space formulas for normalized doubly coprime factorizations for strongly stabilizable and detectable systems. The problem of dynamic compensators is more difficult to resolve and we have obtained only satisfactory results for the class of statically stabilizable systems.

Robustness: Although it has long been believed that strong stability is a very nonrobust property, we have shown that robustness against very general perturbations can be obtained. We have given a parameterization of robustly stabilizing controllers for strongly stabilizable and detectable systems with compact resolvent and finite-dimensional inputs and outputs in the case of coprime factor perturbations. Furthermore, we have shown that the Popov criterion guarantees global asymptotic stability for a feedback system consisting of a strictly positive real, strongly stable system in the forward loop and a sectorial nonlinear perturbation in the feedback loop.

Examples: We have collected a number of examples of physical systems that have a strongly stabilizable and detectable state-space representation. Three of these examples are in our class of dissipative systems with collocated actuators and sensors, and for one of them we have worked out numerically the approximation procedure for the strongly stabilizing solution of the LQ Riccati equation.

10.1   Possibilities for Further Research

Even after the nine previous chapters full of theory, we are still far away from completion of systems and control theory for strongly stabilizable systems. There are still many open problems and possible new research directions. First, there are a number of instances where the results in this book are not yet satisfactory.

Coprime factorizations: We have obtained state-space formulas of normalized doubly coprime factorizations only under the additional assumptions that A has compact resolvent and U and Y are finite-dimensional. It remains an open problem whether the usual formulas also constitute a normalized doubly coprime factorization for the general class of strongly stabilizable and detectable systems. It is to be expected that a solution of this problem of coprime factorizations for more general classes of strongly stabilizable and detectable systems $\Sigma(A, B, C, D)$ will immediately lead to a parameterization of robustly stabilizing controllers for these systems. It is, however, doubtful whether such formulas for normalized doubly coprime factorizations can be found for the general strongly stabilizable/strongly detectable case.

Robust stabilization: Another problem in the parameterization of robustly stabilizing controllers is the fact that our parameterization does not include *all* robustly stabilizing controllers. This deficiency is due to the fact that the usual proofs rely on the construction of a destabilizing perturbation. For this, we need continuity of certain functions on the imaginary axis. As our functions are in $\mathbf{H}_\infty$, we do not have this continuity. The same open problem was observed in Georgiou and Smith [47]. There, the authors expressed their belief in a solution of this problem, although they acknowledged that a completely different approach would be required.

Next, we discuss a number of possibilities for different research directions.

Weak stability: During the development of the current research, an intrinsic problem of working with strong stability showed up at several occasions. All instances had to do with the fact that if A generates the strongly stable semigroup $T(t)$, then it may happen that $T^*(t)$ is not strongly stable. As a consequence, strong stabilizability is not dual to strong detectability and whereas the LQ state feedback does strongly stabilize the system,

LQ output injection may not result in a strongly stable system. A way to avoid this lack of duality would be to concentrate instead on *weak stability*. Let us elaborate on this idea for a moment. We define a system to be weakly stable if it is input stable, output stable, input-output stable and its semigroup is weakly stable. The definitions of weak stabilizability and weak detectability can be derived, in a similar way, from their strong counterparts. In this case, weak stabilizability is the dual of weak detectability. A crucial next step then would be to show whether the application of an arbitrary L_2-input to a weakly stable system leads to a state $z(t)$ that converges to zero weakly (i.e., a generalization of Lemma 2.1.3). We believe that this result holds. With this result, it would probably be possible to extend the theory on the LQ Riccati equation and obtain a result for the filter Riccati equation that is completely dual to the result for the control Riccati equation. Also, the theory for dynamic compensation would not suffer from the lack of duality. The down side is that weak stability is too weak to be useful.

Many of the other results in this book could most likely be extended to weak stability as well. Furthermore, we know from Benchimol [19] that if A generates a weakly stable semigroup $T(t)$ and has compact resolvent, then $T(t)$ is in fact strongly stable. So, for systems whose system operator A has compact resolvent, the results of weak stabilization would even imply strong stability. For certain results, one problem could be that a result analogous to the spectral characterization of strongly stable semigroups by Arendt and Batty [2] is needed. We are not aware of such a characterization of weakly stable semigroups.

Unbounded operators: An important extension of the theory in this book would be to include the possibility of unbounded input and output operators B and C. This would open up an important class of applications, for instance, in problems of structural acoustics (see Banks and Smith [13, 14]. We have never touched upon this problem, but research in this direction has been initiated by Curtain and G. Weiss.

H_∞-control: It would be interesting to see if the state-space approach to H_∞ control, as it was developed for finite-dimensional systems in Doyle et al. [41] and for Pritchard–Salamon systems by van Keulen [104], could be extended to strongly stabilizable systems. Research in this direction is carried out by Staffans [100] and Mikkola [73] in the more general context of well-posed linear systems.

Approximation issues: The approximation procedure for the strongly stabilizing solution of the Riccati equation was presented only for the LQ Riccati equation. It would be very interesting to extend this to more general Riccati equations, like those occurring in H_∞ control. In the context of H_∞ control, the usual procedure is to approximate the system in L_∞ sense by a finite-dimensional system and to do an H_∞ control design for the finite-dimensional plant. If the L_∞ approximation error is sufficiently small, then the closed-loop system is still guaranteed to be stable and there is a certain guaranteed degree of performance for the closed-loop system. This approach is problematic for strongly stable systems because the noncompact Hankel operator causes problems in L_∞ approximation. It seems, therefore, necessary to develop a different approach to the study of approximation for robust control.

Bibliography

[1] V.M. Adamjan, D.Z. Arov and M.G. Krein. Infinite Hankel block matrices and related extension problems. *American Mathematical Society Translations*, 111:133–156, 1978.

[2] W. Arendt and C.J.K. Batty. Tauberian theorems and stability of one-parameter semigroups. *Transactions of the American Mathematical Society*, 306:837–841, 1988.

[3] T. Bailey and J.E. Hubbard, Jr. Distributed piezoelectric polymer active vibration control of a cantilever beam. *AIAA Journal on Guidance, Control and Dynamics*, 8:605–611, 1985.

[4] A.V. Balakrishnan. Strong stabilizability and the steady state Riccati equation. *Applied Mathematics and Optimization*, 7:335–345, 1981.

[5] A.V. Balakrishnan. *A Mathematical Formulation of the SCOLE Control Problem.* Technical Report CR 172581, NASA Langley Research Center, Hampton, VA, 1985.

[6] A.V. Balakrishnan. Stability enhancement of flexible structures by nonlinear boundary-feedback control. In *Boundary Control and Boundary Variations*, Lecture Notes in Control and Information Science 100, pages 18–37, Springer-Verlag, New York, 1998.

[7] A.V. Balakrishnan. Compensator design for stability enhancement with co-located controllers. *IEEE Transactions on Automatic Control*, 36:994–1007, 1991.

[8] A.V. Balakrishnan. Compensator design for stability enhancement with co-located controllers: Explicit solutions. *IEEE Transactions on Automatic Control*, 38:505–507, 1993.

[9] A.V. Balakrishnan. Robust stabilizing compensators for flexible structures with collocated controls. *Journal of Applied Mathematics and Optimization*, 33:35–60, 1996.

[10] J.A. Ball and J.W. Helton. A Beurling-Lax theorem for the Lie group $U(m,n)$ which contains most classical interpolation theory. *Journal of Operator Theory*, 9:107–142, 1983.

[11] H.T. Banks and J.A. Burns. Hereditary control problems: Numerical methods based on averaging approximations. *SIAM Journal on Control and Optimization*, 16:169–208, 1978.

[12] H.T. Banks, G. Propst and R.J. Silcox. A comparison of time domain boundary conditions for acoustic waves in wave guides. *Quarterly of Applied Mathematics*, LIV:249–265, 1996.

[13] H.T. Banks and R.C. Smith. Models for control in smart material structures. In *Identification and Control in Systems Governed by Partial Differential Equations*, H.T. Banks, R.H. Fabiano, and K. Ito, editors, pages 26–44, SIAM, Philadelphia, PA, 1993.

[14] H.T. Banks and R.C. Smith. Feedback control of noise in a 2-D nonlinear structural acoustics model. *Discrete and Continuous Dynamical Systems*, 1:119–149, 1995.

[15] C.J.K. Batty. Tauberian theorems for the Laplace-Stieltjes transform. *Transactions of the American Mathematical Society*, 322:783–804, 1990.

[16] C.J.K. Batty and V.Q. Phóng. Stability of individual elements under one-parameter semigroups. *Transactions of the American Mathematical Society*, 322:805–818, 1990.

[17] J.T. Beale. Spectral properties of an acoustic boundary condition. *Indiana University Mathematics Journal*, 26:895–917, 1976.

[18] C.D. Benchimol. Feedback stabilizability in Hilbert spaces. *Applied Mathematics and Optimization*, 4:225–248, 1977.

[19] C.D. Benchimol. A note on weak stabilizability of contraction semigroups. *SIAM Journal on Control and Optimization*, 16:373–379, 1978.

[20] F. Bucci. Frequency domain stability of nonlinear feedback systems with unbounded input operator. *Dynamics of Continuous, Discrete and Impulsive Systems*, to appear.

[21] F.M. Callier and J. Winkin. LQ-optimal control of infinite-dimensional systems by spectral factorization. *Automatica*, 28:757–770, 1992.

[22] C. Corduneanu. *Integral Equations and Stability of Feedback Systems*. Academic Press, New York, 1973.

[23] R.F. Curtain. Robust stabilizability of normalized coprime factors; the infinite-dimensional case. *International Journal of Control*, 51:1173–1190, 1990.

[24] R.F. Curtain. The Salomon-Weiss class of well-posed linear systems: A survey. *IMA Journal of Mathematical Control and Information*, 14:207–223, 1997.

[25] R.F. Curtain and A. Ichikawa. The Nehari problem for infinite-dimensional systems of parabolic type. *Integral Equations and Operator Theory*, 26:29–45, 1996.

[26] R.F. Curtain and J.C. Oostveen. Absolute stability for collocated systems. In *Proceedings of the International Conference on Optimization: Techniques and Applications (ICOTA)*, Perth, July 1998, L. Cacetta, ed., Curtin University of Technology, pp. 366–372.

[27] R.F. Curtain and J.C. Oostveen. The Nehari problem for nonexponentially stable systems. *Integral Equations and Operator Theory*, 31:307–320, 1998.

[28] R.F. Curtain and J.C. Oostveen. Necessary and sufficient conditions for strong stability of distributed parameter systems. *Systems and Control Letters*, 37:11–18, 1999.

[29] R.F. Curtain and J.C. Oostveen. Normalized coprime factorizations for strongly stabilizable systems. In *Advances in Mathematical Control Theory* (proceedings of a workshop in honour of Diederich Hinrichsen on the occasion of his 60th birthday), U. Helmke, editor, 1999.

[30] R.F. Curtain and J.C. Oostveen. Robustly stabilizing controllers for strongly stabilizable and detectable infinite-dimensional systems. In *Proceedings of the IEEE Conference on Decision and Control*, pages 4274–4290. IEEE Press, Piscataway, NJ, 1999.

[31] R.F. Curtain and A.J. Pritchard. Robust stabilization of infinite-dimensional systems with respect to coprime-factor perturbations. In *Control Theory, Dynamical Systems and Geometry of Dynamics*, K. D. Elworthy, W. N. Everitt and E. B. Lee, editors, pages 437–456. Marcel Dekker, New York, 1993.

[32] R.F. Curtain and A. Ran. Explicit formulas for Hankel norm approximations of infinite-dimensional systems. *Integral Equations and Operator Theory*, 13:455–469, 1989.

[33] R.F. Curtain and B. van Keulen. Robust control with respect to coprime factors of infinite-dimensional positive real systems. *IEEE Transactions on Automatic Control*, 37:868–871, 1992.

[34] R.F. Curtain, G. Weiss and M. Weiss. Coprime factorizations for regular linear systems. *Automatica*, 32:1519–1532, 1996.

[35] R.F. Curtain and H.J. Zwart. The Nehari problem for the Pritchard-Salamon class of infinite-dimensional linear systems: A direct approach. *Integral Equations and Operator Theory*, 18:130–153, 1994.

[36] R.F. Curtain and H.J. Zwart. *An Introduction to Infinite-Dimensional Linear Systems Theory*. Springer-Verlag, New York, 1995.

[37] R.F. Curtain and H.J. Zwart. Riccati equations and normalized coprime factorizations for strongly stabilizable infinite-dimensional systems. *Systems and Control Letters*, 28:11–22, 1996.

[38] E.B. Davies. *One-Parameter Semigroups*. Academic Press, London, 1980.

[39] C.A. Desoer and M. Vidyasagar. *Feedback Systems: Input-Output Properties*. Academic Press, New York, 1975.

[40] A. Devinatz and M. Shinbrot. General Wiener-Hopf operators. *Transactions of the American Mathematical Society*, 145:467–494, 1969.

[41] J.C. Doyle, K. Glover, P.P. Khorgonekar and B.A. Francis. State-space solutions to standard H_2 and H_∞ control problems. *IEEE Transactions on Automatic Control*, 34:831–847, 1989.

[42] P.L. Duren. *Theory of H_p Spaces*. Academic Press, New York, 1970.

[43] H. Dym, T.T. Georgiou and M.C. Smith. Explicit formulas for optimally robust controllers for delay systems. *IEEE Transactions on Automatic Control*, 40:656–669, 1995.

[44] C. Foias and A. Tannenbaum. On the Nehari problem for a certain class of L_∞-functions appearing in control theory. *Journal of Functional Analysis*, 74:146–159, 1987.

[45] P.A. Fuhrmann. On the Corona theorem and its application to spectral problems in Hilbert space. *Transactions of the American Mathematical Society*, 132:493–520, 1968.

[46] P.A. Fuhrmann. *Linear Systems and Operators in Hilbert Space*. McGraw–Hill, New York, 1981.

[47] T.T. Georgiou and M.C. Smith. Optimal robustness in the gap metric. *IEEE Transactions on Automatic Control*, 35:673–685, 1990.

[48] T.T. Georgiou and M.C. Smith. Robust stabilization in the gap metric: Controller design for distributed plants. *IEEE Transactions on Automatic Control*, 37:1133–1143, 1992.

[49] J.S. Gibson. Linear-quadratic optimal control of hereditary differential systems: Infinite-dimensional Riccati equations and numerical approximations. *SIAM Journal on Control and Optimization*, 21:95–139, 1983.

[50] K. Glover, R.F. Curtain and J.R. Partington. Realisation and approximation of linear infinite-dimensional systems with error bounds. *SIAM Journal on Control and Optimization*, 26:863–898, 1988.

[51] P. Grabowski. On the spectral-Lyapunov approach to parametric optimization of distributed parameter systems. *IMA Journal of Mathematical Control and Information*, 7:317–338, 1990.

[52] M. Green, K. Glover, D. Limebeer and J. Doyle. A J-spectral factorization approach to $\mathcal{H}_\infty$ control. *SIAM Journal on Control and Optimization*, 28:1350–1371, 1990.

[53] S. Hansen and G. Weiss. New results on the operator Carleson measure criterion. *IMA Journal of Mathematical Control and Information*, 14:3–32, 1997.

[54] F. Huang. Strong asymptotic stability of linear dynamical systems in Banach spaces. *Journal of Differential Equations*, 104:307–324, 1993.

[55] K. Ito. Strong convergence and convergence rates of approximating solutions for algebraic Riccati equations in Hilbert spaces. In *Distributed Parameter Systems*, F. Kappel, K. Kunisch and W. Schappacher, editors, pages 151–166, Springer-Verlag, New York, 1987.

[56] K. Ito and G. Propst. Legendre-tau-Padé approximations to the one-dimensional wave equation with boundary oscillators. *Numerical Functional Analysis and Optimization*, 90:57–70, 1997.

[57] S.M. Joshi. *Control of Large Flexible Space Structures*, Lecture Notes in Control and Information Science 131. Springer-Verlag, Berlin, 1989.

[58] F. Kappel and D. Salamon. An approximation theorem for the algebraic Riccati equation. *SIAM Journal on Control and Optimization*, 28:1136–1147, 1990.

[59] T. Kato. *Perturbation Theory of Linear Operators*. Springer-Verlag, New York, 1966.

[60] V. Kučera. New results in state estimation and regulation. *Automatica*, 17:745–748, 1981.

[61] J.P. Lasalle. *The Stability of Dynamical Systems*. SIAM, Philadelphia, PA, 1976.

[62] N. Levan. The stabilizability problem: A Hilbert space operator decomposition approach. *IEEE Transactions on Circuits and Systems*, 25:721–727, 1978.

[63] N. Levan. On some relationships between the LaSalle invariance principle and the Nagy-Foias canonical decomposition. *Journal of Mathematical Analysis and Applications*, 77:493–504, 1980.

[64] N. Levan. Approximate stabilizability via the algebraic Riccati equation. *SIAM Journal on Control and Optimization*, 23:153–160, 1985.

[65] N. Levan. Strong stability of quasi-affine transforms of contraction semigroups and the steady-state Riccati equation. *Journal of Optimization Theory and Applications*, 45:397–406, 1985.

[66] N. Levan. Stabilization by the feedbacks $-B^*$ and $-B^*P$. In *System Modelling and Optimization*, Lecture Notes in Control and Information Science 84, Springer-Verlag, New York, pages 542–550. 1986.

[67] N. Levan. Stability enhancement by state feedback. In *Control Problems for Systems Described by Partial Differential Equations and Applications*, pages 313–324. Springer-Verlag, Berlin, 1987.

[68] N. Levan and L. Rigby. Strong stabilizability of linear contractive control systems on Hilbert space. *SIAM Journal on Control and Optimization*, 17:23–35, 1979.

[69] J.L. Lions. *Optimal Control of Systems Governed by Partial Differential Equations*. Springer-Verlag, Berlin, 1971.

[70] H. Logemann, E.P. Ryan and S. Townley. Integral control of infinite-dimensional linear systems subject to input saturation. *SIAM Journal on Control and Optimization*, 36:1940-1961, 1998.

[71] J.M. Maciejowski. *Multivariable Feedback Design*. Addison-Wesley, Reading, MA, 1989.

[72] D.C. McFarlane and K. Glover. *Robust Controller Design using Normalized Coprime Factor Plant Descriptions*, Lecture Notes in Control and Information Sciences 138. Springer-Verlag, New York, 1989.

[73] K. Mikkola. *On the Stable H_2 and H_∞ Infinite-Dimensional Regular Problems and their Algebraic Riccati Equations*. Technical Research Report A383, Helsinki University of Technology, Institute of Mathematics, Helsinki, Finland, 1997.

[74] B.P. Molinari. Equivalence relations for the algebraic Riccati equation. *SIAM Journal on Control*, 11:272–285, 1973.

[75] Ö. Morgül. Dynamic boundary control of a Euler-Bernoulli beam. *IEEE Transactions on Automatic Control*, 37:639–642, 1992.

[76] Ö. Morgül. Dynamic boundary control of the Timoshenko beam. *Automatica*, 28:1255–1260, 1992.

[77] Ö. Morgül. A dynamic control law for the wave equation. *Automatica*, 30:1785–1792, 1994.

[78] P.M. Morse and K.U. Ingard. *Theoretical Acoustics*. McGraw–Hill, New York, 1968.

[79] K.S. Narendra and J.H. Taylor. *Frequency Domain Criteria for Absolute Stability*. Academic Press, New York, 1973.

[80] J.C. Oostveen and R.F. Curtain. Riccati equations for strongly stabilizable bounded linear systems. *Automatica*, 34:953–967, 1998.

[81] J.C. Oostveen and R.F. Curtain. Robustly stabilizing controllers for dissipative infinite-dimensional systems with collocated actuators and sensors. *Automatica*, 36:337–348, 2000.

[82] J.C. Oostveen and R.F. Curtain. The Popov criterion for strongly stable distributed parameter systems. *International Journal of Control*. Submitted for publication.

[83] J.C. Oostveen, R.F. Curtain and K. Ito. An approximation theory for strongly stabilizing solutions to the operator LQ Riccati equation. *SIAM Journal on Control and Optimization*, to appear.

[84] J.R. Partington. *An Introduction to Hankel Operators*. London Mathematical Society Student Texts. Cambridge University Press, Cambridge, UK, 1988.

[85] A. Pazy. *Semigroups of Linear Operators and Applications to Partial Differential Equations*. Springer-Verlag, New York, 1983.

[86] V.M. Popov. Absolute stability of nonlinear systems of automatic control. *Automation and Remote Control*, 22:857–875, 1961.

[87] S.C. Power. *Hankel Operators on Hilbert Space*. Pitman, Boston, MA, 1982.

[88] A. Ran. Hankel norm approximation for infinite-dimensional systems and Wiener-Hopf factorization. In *Modelling Robustness and Sensitivity Reduction in Control Systems*, R.F. Curtain, editor, NATO ASI Series, pages 57–70. Springer-Verlag, New York, 1986.

[89] R. Rebarber. Conditions for the equivalence of internal and external stability for distributed parameter systems. *IEEE Transactions on Automatic Control*, 38:994–998, 1993.

[90] M. Rosenblum and J. Rovnyak. *Hardy Classes and Operator Theory*. Oxford University Press, New York, 1985.

[91] D.L. Russell. Linear stabilization of the linear oscillator in Hilbert space. *Journal of Mathematical Analysis and Applications*, 25:663–675, 1969.

[92] M. Slemrod. The linear stabilization problem in Hilbert space. *Journal of Functional Analysis*, 11:334–345, 1972.

[93] M. Slemrod. A note on complete controllability and stabilizability for linear control systems in Hilbert space. *SIAM Journal on Control*, 12:500–508, 1974.

[94] M. Slemrod. Feedback stabilization of a linear control system in Hilbert space with an a priori bounded control. *Mathematics of Control, Signals and Systems*, 2:265–285, 1989.

[95] M.C. Smith. On stabilization and the existence of coprime factorizations. *IEEE Transactions on Automatic Control*, 34:1005–1007, 1989.

[96] O.J. Staffans. Quadratic optimal control of stable systems through spectral factorization. *Mathematics of Control, Signals and Systems*, 8:167–197, 1995.

[97] O.J. Staffans. Quadratic optimal control of stable well-posed linear systems. *Transactions of the American Mathematical Society*, 349:3679–3716, 1997.

[98] O.J. Staffans. Coprime factorizations and well-posed linear systems. *SIAM Journal on Control and Optimization*, 36:1268–1292, 1998.

[99] O.J. Staffans. Feedback representations of critical controls for well-posed linear systems. *International Journal of Robust and Nonlinear Control*, 8:1189–1217, 1998.

[100] O.J. Staffans. On the distributed stable full information H_∞ problem. *International Journal of Robust and Nonlinear Control*, 8:1255–1305, 1998.

[101] O.J. Staffans. Quadratic optimal control of well-posed linear systems. *SIAM Journal on Control and Optimization*, 37:131–164, 1998.

[102] S.R. Treil. The theorem of Adamjan-Arov-Krein: A vector variant. *Zap. Nauchn. Semin. Leningrad. Otdel. Math. Inst. Steklov (LOMI)*, 141:56–71, 1985 (in Russian).

[103] B. van Keulen. Hankel operators for non-exponentially stabilizable infinite-dimensional systems. *Systems and Control Letters*, 15:221–226, 1990.

[104] B. van Keulen. H_∞-*Control for Distributed Parameter Systems: A State-space Approach*. Birkhäuser, Boston, MA, 1993.

[105] M. Vidyasagar. *Nonlinear Systems Analysis*, second edition. Prentice-Hall, Englewood Cliffs, NJ, 1993.

[106] G. Weiss. Admissibility of unbounded control operators. *SIAM Journal on Control and Optimization*, 27:527–545, 1989.

[107] G. Weiss. Admissible observation operators for linear semigroups. *Israel Journal of Mathematics*, 65:17–43, 1989.

[108] G. Weiss. The representation of regular linear systems on Hilbert spaces. In *Control and Estimation of Distributed Parameter Systems*, W. Schappacher, F. Kappel and K. Kunisch, editors, pages 401–416. Birkhäuser-Verlag, Basel, Switzerland, 1989.

[109] G. Weiss. Representation of shift-invariant operators on L_2 by H_∞ transfer functions: An elementary proof, a generalization to L_p and a counterexample for L_∞. *Mathematics of Control, Signals and Systems*, 4:193–203, 1991.

[110] G. Weiss. Regular linear systems with feedback. *Mathematics of Control, Signals and Systems*, 7:23–57, 1994.

[111] M. Weiss. *Riccati Equations in Hilbert Spaces: A Popov function approach*. PhD thesis, Rijksuniversiteit Groningen, The Netherlands, 1994.

[112] M. Weiss. Riccati equation theory for Pritchard-Salamon systems: A Popov function approach. *IMA Journal of Mathematical Control and Information*, 14:45–84, 1997.

[113] G. Weiss and R.F. Curtain, Stabilizing controllers with internal loop for infinite-dimensional systems. In *Proceedings of the European Control Conference*, pages 3294–3297, 1995.

[114] M. Weiss and G. Weiss. Optimal control of stable weakly regular linear systems. *Mathematics of Control, Signals and Systems*, 10:287–330, 1997.

[115] D. Wexler. Frequency domain stability for a class of equations arising in reactor dynamics. *SIAM Journal on Mathematical Analysis*, 10:118–138, 1979.

[116] D. Wexler. On frequency domain stability for evolution equations in Hilbert spaces via the algebraic Riccati equation. *SIAM Journal on Mathematical Analysis*, 11:969–983, 1980.

[117] J.C. Willems. Dissipative dynamical systems. Part I: General theory. *Archive for Rational Mechanics and Analysis*, 45:321–351, 1972.

[118] J.C. Willems. Dissipative dynamical systems. Part II: Linear systems with quadratic supply rates. *Archive for Rational Mechanics and Analysis*, 45:352–393, 1972.

[119] J.L. Willems. *Stability Theory of Dynamical Systems*. Nelson, London, 1970.

[120] Y. You. Dynamical boundary control of two-dimensional Petrovsky system: Vibrating rectangular plate. In *Analysis and Optimization of Systems*, A. Bensoussan and J.L. Lions, editors, Lecture Notes in Control and Information Sciences 111, pages 519–530. Springer-Verlag, Heidelberg, 1988.

Index